I WISH YOU...

I WISH YOU...

FERUZA SHAH MURAD

authorHOUSE®

AuthorHouse™ UK
1663 Liberty Drive
Bloomington, IN 47403 USA
www.authorhouse.co.uk
Phone: 0800.197.4150

Scripture quotations are from the Holy Bible, King James Version (Authorized
Version). First published in 1611. Quoted from the KJV Classic Reference
Bible, Copyright © 1983 by The Zondervan Corporation.

Published by AuthorHouse 03/27/2015

ISBN: 978-1-5049-3907-2 (sc)
ISBN: 978-1-5049-3908-9 (e)

Print information available on the last page.

Contents

I'm devoting this book to all my teachers- to everyone I came into the contact with. Majority of my teachers are thousands of my patients...

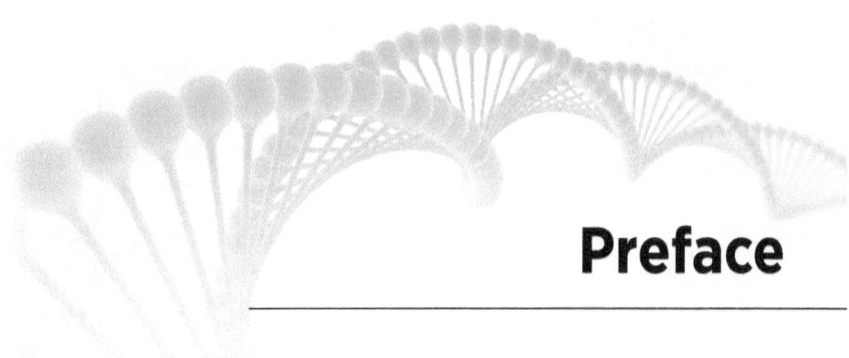

Preface

I'm a qualified medical doctor. Nowadays I'm working in Central London with the team of stellar consultants, wittingly wise professors and brilliant nurses. I offer wide range of diagnostics and treatments in leading London Hospitals for my patients and have a network of patients with their families from around the world and indeed I want to help all of them to attain better health.

The purpose of this book is to help my patients and everyone who wants make best of their lives.

In recent years, ground breaking, novel treatments and improved old techniques of treatments made possible to fight diseases that deemed to be incurable in past, better medicines allows to live normal life with serious conditions and the medicine evolves further. With the modern technology advance, with overwhelm of information that bombards our brains, new problems, which lead to the range of diseases emerge.

But why certain treatments would heal some individuals and make no difference to others?

Don't we all have approximately same anatomy and same structure of the tissues, cells that construct our organs? Isn't that only the DNA and hence inherited illnesses and conditions make us such unique that we respond to the treatment in an unprecedented manner sometimes? Why we responding differently to the same treatment from the same disease and why there are lots of similar patients who do not gain any benefit from the treatments which perfectly work for others?

How is important for us to know what our world and our bodies consist from and how do they function, how do they interfere?

We think that our lives and hence destiny depend on external factors, those like our financial needs, relationships and conditions. But modern science tells us that it is not a necessary trend. In fact, our health, our environment, our social and financial status shaped by our inner world, by our Self.

While some circumstances are beyond our control and we are unable to change them in past and in present, it is our responsibility to revise our conditions and try to improve them for future for better.

As soon as we become aware things don't seem go our way, or when we find ourselves unhappy at that point we are to shift out navigation towards the success, towards the perfect health, to enjoy life fully in it's miraculous beauty.

I would refer the state of unhappiness to the point, when we need to rethink our lives, to develop our awareness, to use our knowledge and head to the perfect health, harmonious life and create happiness around us. In other words that would be the point of our nadir – the point of the beginning of success story, of our rise. The purpose of the state of unhappiness is to think, to judge, to weight, to develop new ideas and probabilities, to evolve.

I want to extend the concept on how it may work for you in this book.

But how to use the knowledge from this book depends on how quick you want to achieve your dream, if you have one or many.

This knowledge comes to everyone with their life experience, but perhaps if we know earlier, we are more likely to use this knowledge throughout our journey and navigate correctly our lives, avoiding pitfalls and occurrence of bad events.

While I accept that the knowledge I'm offering to the reader and it's impact on our lives and health is only a theory, I am confident that conducted researches in the area that regarded as the metaphysics today, tomorrow will prove it's true. It will change current approaches in conventional medicine, which is nowadays improving significantly towards integrative medicine but indeed is yet to become perfect.

Until then I want to contribute in promotion of the mental health and emotional state, as I personally believe that every disease, including congenital ones, originate from the state of our mind and from The Mind.

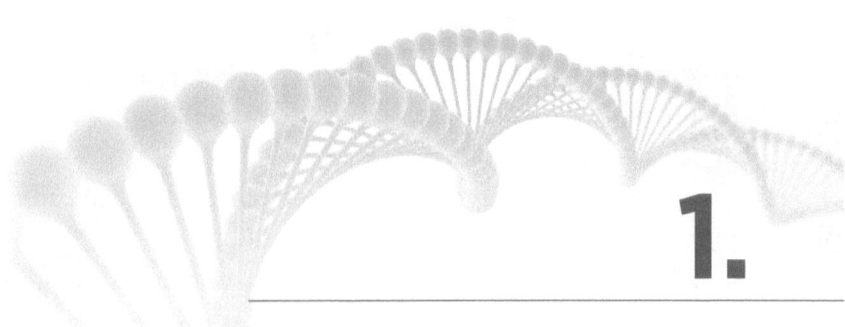

1.

The Beginning

In the beginning, there was the Word.

In what form? Told loudly, or whispered out? Or just in a form of the thought, an idea, or a wish?

Word, spelled out is a sound wave. Not pronounced, it is a thought. In some stage, particularly when replayed, thoughts become intentions, more powerful then the mere thought or the word. Intention transforms the thought into the action. Wish transforms the word, the thought to the intention. Have you ever thought about the wish? Where do our wishes coming from?

It sounds weird in 21st century but in fact it is true that nobody really knows how exactly things and events occur in the lives of individuals neither do we know the exact pathogenesis of the illnesses we face. That is not only because the factors contributing into the diseases are miscellaneous.

Modern medicine acknowledges those factors such as lifestyle, diet, genetic (inherited) predisposition, autoimmune, infections are the major culprits. But all we know or can explain is only a part of the causes of the diseases. For example, we cannot explain how autoimmune processes get sparkled or why 3:1 risk ratio of developing cancer would affect some members of the same family and wouldn't cause any harm to others.

Sometimes it seems our ancient ancestors knew even more than we do. Somewhere deep inside I have an intuitive feeling that everybody has: life events with the circumstances we face throughout our lives doesn't happen by the mere coincidence.

There are almost always some sort of the mechanisms that lead to the certain scenario in our lives and health conditions.

While it is impossible to classify the events and circumstances that lead people to the similar health conditions, there are identified predispositions that can be summarized in consistency leading us to conclusions what are the risk factors for the certain diseases.

For example, healthy diet and physical activity are the key approaches to avoid the stroke or diabetes.

But that is not enough, as the evidence show.

Healthy, sportive people can have strokes; young children and babies may have strokes, too.

There are interesting publications and articles regarding the stroke statistics, telling us that "a quarter of strokes are in working –age people and children and babies also have strokes" - (Dr Clare Walton, Stroke Association, BBC news 10 January 2013 health report by J. Gallagher.)

What is the modern science says is that avoiding the risk factors we aware of, like unhealthy lifestyle and diet in order to achieve better health only is not enough. The stress, the major contributing factor of almost every disease, needs to be avoided. Is there any feasible tactics to avoid the stress in modern life? Indeed there are, providing that you re-think, re- shape, re- decide on your current state of your mind. If you do, you achieve the certain potential that every person does have. These potentials, if misused, harm us.

There are states of the mind, emotional background, better moods, positive thinking factors that empower our immune system and help us to avoid certain diseases and even to keep our genes carrying grief diseases inactivated.

I was always wondering: How does that work? In my view, a positive thinking depends on our emotional state and only in controlled emotional state we can manage our thoughts. Our environment mainly affects our emotional state, by the nature of frequencies we are tuned to. Yes, the frequencies are the main affecting factor of our emotional state.

I'll explain the nature of frequencies I'm talking about. Mainly, these are the physical properties-the frequencies, of our thoughts.

The best way to keep control of our thoughts varies from person to person, but in my view and from my experience, it is a daydreaming. Most of the patients I faced, who'd been successfully treated by the multidisciplinary teams, were tuned positive. They are great daydreamers in their majority. I'd compare the daydreaming to the philosophy, to the higher thinking. We all own such ability – to daydream, be we a cleaner, an academician, a royalty or a soldier.

Having a dream is a great power; it is a main force of the evolution of everything in the Universe. Dream is the natural force of evolution. Dreaming moves us forward, when it flourishes within us, but if stagnated, causes damage to our personality.

During the course of our evolution, mainly in our adulthood, we find ourselves unhappy, unsatisfied, disappointed, hurt due to the miscellaneous reasons.

Depression is the main curse of the modern society, and it's elements lie at the bottom of pathogenesis of every disease. To come out from low mood, from the depression, to attain perfect health, lasting youth and anything else you might want, you need to know the basics of our existence and interference with the environment.

When we know how things function in our world, we can influence our life and body states by using natural effects, which work irrespective of our opinion about them.

We can do so by knowing, remembering and reminding ourselves the physics of our thoughts.

Whether you want it or not, accept that or deny, where ever in the world you live, you are under the influence of natural laws that rule the whole Universe.

Knowing their basics will help you to navigate towards your goals.

We are nowadays approaching a brand new world, where the recognition of the fact that subtle forces manage the Universe not only work as a single feasible theory, but will be a proven axiom, or the fundamental truth and is used towards our progressive future.

While invisible to the naked eye, such forces do exist. They are fundamental basement of our existence and they make the Universe functioning in its way.

So, it worth always to remember: in the beginning, there was the Word.

What is the most important: the thought or the word? Well, it is an old and funny dilemma like if there was first a chicken or an egg? Of course, there is a little difference between the thoughts and the words. But word (and hence, the thought) has such a power that has brought up the whole Universe into existence. It was primordial word – the thought that carried such mighty power.

Written word would remind me a structure of the DNA, if you write many words together and twist them around.

DNA is an information carrier, storage we've inherited from our family and combination of the features unique to us - our genes.

They activate or stay inactivated throughout our lives. Some of them would carry on our unique talents, the others our physical appearance features, like a color of our eyes, hair structure while others would be responsible for the disease.

Encrypted, they lie in the core of every cell of our bodies. Each cell, every creature, has own DNA.

The DNA consists of the molecules, and molecules (like any other molecules) composed from the atoms.

Atoms made up from the tiny quarks – the bits of information, broadly equal to the thoughts, which are bits of information, too.

Quarks are the tiny particles, merely visible; they appear and disappear under lenses of sophisticated microscopes.

So, they are wave –like particles, they exist in a wave fashion, but at the same time they exhibit a property of duality, they oscillate, they appear and disappear at the same time.

Some scientists say: they simply don't exist, the tiny quarks, the building material of everything existing in the Universe exist in our imagination only!

The other science giants, mainly supporters of the Quantum field theory, which is nowadays widely accepted model of fundamental physics state that the quarks are oscillating in the material reality, but where do they come from and where do they disappear?

Are the scientists claiming the quarks exist in our imagination only are saying a nonsense or are they saying the truth? Or what the other quantum physicists say us that quarks are that much subtle that no attention they deserve?

I'll tell you my understanding of quarks existence later, but let return to the beginning, where there was a word, an idea, a thought.

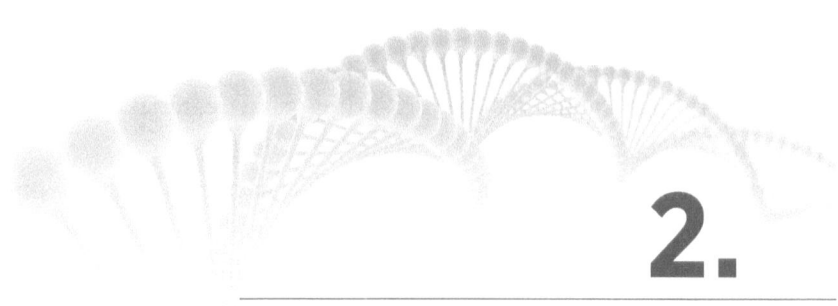

2.

The Word

In any form I believe, the word is a sound wave and complies with the nature of waves physics.

Let's think about my statement: Word, spelled out is the sound wave but if not pronounced, is an idea, a thought.

Non- pronounced word is the sound wave, too. It's not my only hypothesis; there are plenty of scientists who teach us that the thoughts are the waves.

Marco Bischoff, for example, one of the prominent biophysicists of our time, reckons that they are waves, like the radio waves.

According to Marco Bischoff, whom I 've had an honor to talk to during the break of his lecture in London in 2012, thoughts resemble to the radio waves but they travel in space and time. We catch them, as antennae, if we are tuned on those frequencies.

When he told us that, I immediately recalled the phrase from the Quran: "Every soul finds what it sought", but as my religion expands to the Science, and I believe that the Quran, the Bible and the Tarot are the purely Scientific books and should only be interpreted by the Scientists rather then by ill-educated fanatics, I'd get your attention back to the Science.

Marco Bischoff is the President of the International Institute of Biophysics and well known for his contributions in biophotons research (See the ref. at MarcoBischoff.com).

Biophotons are the tiny electromagnetic waves, or the light particles, which every live organism emits, and the Russian scientist Alexander G. Gurvich has first discovered them in 1923.

Rediscovered in 1970s by the group of European and American scientists, they have been proved to originate from the DNA, by German biophysicist Fritz-Albert Popp in 1974.

Theoretically, the biophotons may control neuro- humoral and other biochemical exchanges in the organisms, which constantly occur in live bodies.

Here is the interesting puzzle for the medical practitioner.

If the biophotons are emitted by the DNA, and the DNA is encrypted information of our organisms, which is very much resemble to a written world, and words are the thoughts, then who do writes that words?

Every DNA is an information carrier. It's very well known that we do inherit our DNA from our parents and family. Who decides what bits of the DNA to encrypt in our children? Why the DNA emits light? Is light wave is similar to the thought waves or just functions like an antennae? What affects our DNA alleles that make them activated in adult life or keeps them dormant?

Every non- pronounced word in mind or on paper is a thought.

Thoughts are a sort of sound waves. That sound is much beyond the hearing range. I believe, a sort of the supra-sonic waves. May be, they'd be named as an infra- sonic wave in a future. But what I certainly believe in that the thoughts are the sound waves.

Every quark is a thought.

If that what modern quantum physics say us the truth, every quark is a bit of information, I would refer every quantum to the thought.

Every thought becomes manifest in time, whatever absurd it might be. Every thought oscillates from non- material, imagination reality to the material reality. Imagination is a form of reality. I belief that imagination is also a form of hypnosis in some extend. We all have our own imagination. Albert Einstein told once: Imagination is EVERYTHING. I also think that everything is hypnosis, too.

7

Therefore, the thoughts also can hypnotize. The power and speed of manifestation of your thoughts into your physical reality depends on the intensity you replay your thoughts in your mind and on desire you put into your thoughts. Desire depends on the incentive, stimuli. Idea is hypnosis. We all at different stages of our evolution and in different stages of our personal life journeys are hypnotized by the ideas, whether we develop them ourselves, or picking them from the others.

Every thought is a wave, a sound wave. Their dimension so tiny, that the normal ear and brain preceptors cannot hear them, unless they originate from our brain, given us in certain (hypnotic, night dreaming) time, or in form of the idea.

Since their frequencies that tiny, they travel through space and time.

That is a universal feature of the sound waves: the higher the frequencies, the shallower the penetration to the media, the lower their frequencies, the farer, the deeper inside the media they travel.

I'll explain why intuitively I assume thoughts are the waves that travel through space and time, not only in media, and how thoughts of others may impact us, if we do not control our "antenna", - our brain in a certain ways.

By the way, I believe that we have our "antenna" that catches the thoughts as Marco Bishoff suggests, in form of physical cells in our brain. They are similar to the cells, discovered by the scientists from the University College (London 2014) in rodents that navigate them in space. They call them "the brain GPS cells ". I believe we also have the cells in our brain that act like a cords of the musical instrument and they produce our own thoughts. If replayed, those thoughts become into our awareness and then may turn into the intention or into the words.

The thoughts emitted into the media are the sound waves released around us and travel in space and time, oscillating the molecules of the media, creating vibrations. But having said that, I don't know whether the vibration of the material media more essential then the vibration of the non-material reality.

Those vibrations caused by our thought in the material media may have negligible measures but in non-material reality, in imagination of

The God, in The Mind, they might have similar or even greater power then in the material world, in other words, greater than the physical exhibition of the thoughts in non-physical environment. Yet again, the physics of the thoughts, we may get trapped in by hypnosis of their frequencies and wavelength is as simple as it complex.

Imagine someone who lives a life of a prosperous, successful person. Such people know the physics of the frequencies. They know, if trapped in wrong frequencies, one cannot get rid of them without professional help. They know the bad frequencies start when you allow them in to your life. Your health, finance, relationships, all can be affected if you loose control and allow yourself to fall into the stress frequencies. They do not allow the stress into their lives, but even if the stress occurs, they do not fall into the hypnoses of bad frequencies. They do not spend too much time by analysing the stress causes, neither bother by blaming others.

Simply let go your thoughts about the stress don't let the stress entrap you in, keep going on. There are always things you wanted to achieve, switch to those things, do them now, and let bad frequencies to pass by.

One of my great teachers whom I really appreciate as one of the wisest men in the world, Professor S. Lingam, once very wittily told me when I was complaining on issues I have had with one of the most weird staff personal: I don't have any energy to fight on this person's stupidity.

That was an extremely precise formulation: to not waste the energy for useless fight! In some person, healthy debates can bring to the constructive solutions but in others, we know the fight will not bring any fruit. That is an issue for the person who has been trapped into the bad frequencies and remains under hypnosis, within the whirlpool of their delusions. That person might need professional help, but not a waste of your energy, that simply wouldn't work!

We all had the situations, where our ideas seemed get picked up by someone else, and we often even regret, that didn't put our ideas first somewhere noticed.

We even blame others in stealing of our ideas, literally and as a figurative expression. But that does indicate for me that the thoughts are waves and travel beyond the space- time dimension.

Every thought is a sound wave, every non-pronounced word is a thought, every thought has it's own frequency, wavelength and other features of the physical sound wave in different extend. Words allow us to the reality, peculiar to those words. So, to create a better reality, one must carefully choose the words in their minds as the every word makes us approach the reality of that word. Simply to be wealthy, happy, healthy, we need in first instance to choose the words of health, wealth, prosperity, and beauty to animate those frequencies of relative waves around us and within our brain. Rest assured, our brain will found ways to animate every word in our mind, and the only matter is the Time.

Since we are developing the notion of the word, I would refer the written word to the equality of our DNA.

Let's think of the DNA as a written word. There is a need to understand the power of the written word. Every psychology therapist advises to their patients to keep a diary, to write good wishes into the diary and see their manifestation. How this phenomena works, the science is yet to reveal. I would add, that by writing the letters, writing a diary, you would literally embellish your desires into your DNA.

Think of your DNA structure you'll pass to your children.

I'm going to develop this idea further but first …

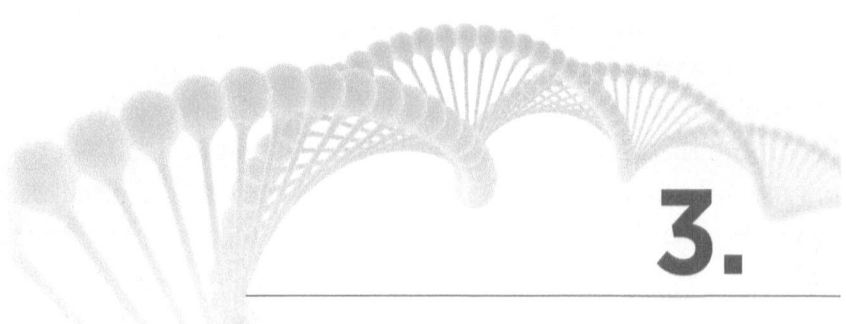

3.

The Ultrasound

Let me explain you what is the ultrasound.

The ultrasound wave is a wave of sound beyond the hearing threshold.

Nowadays, we can measure the length, frequency and speed among the other physical properties of these sound waves and apply the magic of the ultrasound to many areas of medicine and technology.

I use the ultrasound in a daily practice as a diagnostic tool, and can tell what is going on in the organs of the human body, without my patient feeling those waves, that travel through their body. The patient feels only a touch of the transducer to their skin, but not the sound waves.

We can't hear that sound waves, can't feel them, neither we can see them.

But Sound waves are there. They emitted by the transducer, permeate the skin, the underlying tissues, travel trough the organs and at the same time are reflected by the tissues and organs. The computer then transforms the reflections of the sound waves to the screen images.

Organs reflect sound waves with different intensity unique to the type of the tissue.

For instance, liquids like a fresh blood, or gall in the gallbladder, or urine in the bladder will reflect the waves with less or no intensity unlike the solid tissues, like a kidney, or liver tissue, which appear brighter then the liquids on the screen.

Remember: the sound waves beyond the hearing range (we cannot hear them, that tiny they are) called an ultrasound waves. It is a very old and well-known notion. We cannot hear them, but they are physical waves. We use those waves for the diagnostic procedures and for the treatment.

Is that harmful? No more then a sound of the bat, which flied along.

But, certainly, all tissues and organs and hence every cell have an ability, or property, to receive the sound /the ultrasound wave, to adsorb it (like the bone tissue does) to transmit, and who knows? may have ability to process them and, the main property of every tissue and hence every cell, reflect the sound waves with the different intensity …

The non- organic matter does reflect the sound waves, too. In the mountains our echo reflected by the rocky surface is an example.

Here what the Wikipedia tells us about the Sound.

Sound can propagate through compressible media such as air, water and solids as longitudinal waves and also as a transverse waves in solids. The sound waves are generated by a sound source, such as the vibrating diaphragm of a stereo speaker. The sound source creates vibrations in the surrounding medium. As the source continues to vibrate the medium, the vibrations propagate away from the source at the speed of sound, thus forming the sound wave. At a fixed distance from the source, the pressure, velocity, and displacement of the medium vary in time. At an instant in time, the pressure, velocity, and displacement vary in space. …During propagation, waves can be reflected, refracted, or attenuated by the medium.[4]

- The propagation (of the sound – author) is also affected by the motion of the medium itself. For example, sound moving through wind. Independent of the motion of sound through the medium, if the medium is moving, the sound is further transported.

When sound is moving through a medium that does not have constant physical properties, it may be refracted (either dispersed or focused).[4]

Spherical compression (longitudinal) waves

From Wikipedia, the free encyclopedia

The mechanical vibrations that can be interpreted as sound are able to travel through all forms of matter: gases, liquids, solids, and plasmas. The matter that supports the sound is called the medium.

Wikipedia

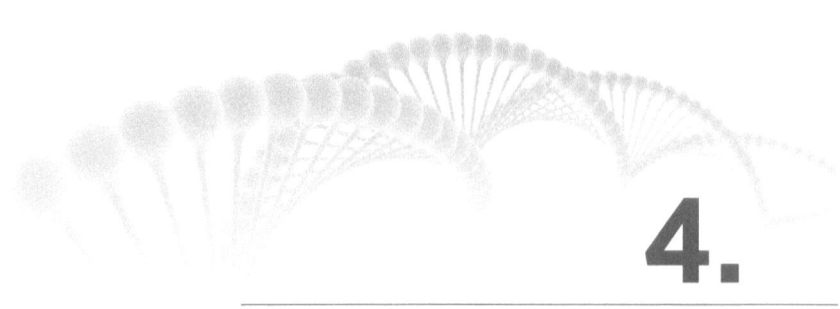

4.

Vibrations

I believe we all exist in One Reality. It is also an established scientific theory, widely and formally accepted by the science community throughout the world. That One Reality, which is beyond space-time gravity dimension, doesn't comply with the rules and physics of those dimensions but is the source of those dimensions and everything that exists and complies with them.

Some nowadays call that Reality the Quantum Field; I call it The Supreme Imagination. That is an only Reality that rules the Universe. When we imagine, when we dream, we then are similar to the Supreme Imagination, as then we create. We connect to it by mere thinking. The fact is that we think even when we sleep. Yet we exist within its imagination.

It is a Reality, where subtle forces in their mean, existence and actions are much more powerful then those forces that exhibit their properties, physically measurable and can be observed. That is a Reality which embrace every single cell, every atom and every particle of it and the entire Universe with everything that exist beyond it.

It is full of vibrations, created by thoughts and by wishes.

In realm of that Reality, thoughts are much more essential, then the words; intentions mean more then actions.

When you remind yourself that in Real Reality your thoughts mean more then your words, intentions more then your actions, you will choose the right thoughts and intentions for yourself, for those whom you love as your every thought you replay in your mind, every wish you entertain in your heart directly affects your life, your reality and of those whom you love, whom with you share your DNA.

It is my belief that good or bad, thoughts and wishes, once released, are distributed outside. They do not disappear, according to law of energy perdurability (energy conservation). They accumulate according to their length/weight /frequency in space we yet to explore. They are sound waves, that create vibrations and if accumulated in space or in Immeasurable Reality, then the tales about the haven and the hell are quite feasible theory.

If single word created the whole Universe, thoughts and wishes of billions people will definitely create accumulations of those wishes and thus eternal stores of thoughts. It is logical to assume, that negative thoughts would have same length/frequencies/measures and positive ones would have same ones.

One day we will be able to measure our thoughts sound waves and tell with the confidence, where are the positive ones are accumulated and where are do the negative ones.

My view of the thoughts physics in physical word is that they are sound waves of very small, even negligible dimensions.

I want to present my view on how they affect our bodies and our life circles.

Sound waves, while propagate the media (let's assume our brain in first instance) vibrate the particles of the media...Wave motion transfers the energy from one point to other.

Let's read what Wikipedia tells us about the waves physics:

> "There are two main types of waves. Mechanical waves propagate through a medium, and the substance of this medium is deformed. The deformation reverses itself owing to restoring forces resulting from its deformation. For example, sound waves propagate via air molecules colliding

with their neighbors. When air molecules collide, they also bounce away from each other (a restoring force). This keeps the molecules from continuing to travel in the direction of the wave." –Wikipedia

Think about our body cells, that this tiny thoughts /sound waves propagate through and if they can change those cells. Think about our DNA, as you know, it consists of the molecules that attached each to other in some sort of order. If constantly similar thoughts –sound waves propagate through them, then, it is no wander they activate certain genes while keeping others inactivated. I do not mention miscellaneous neuro- humoral changes that our emotions cause in our bodies, but remember, the thoughts trigger our emotions.

"The second main type of wave, electromagnetic waves, do not require a medium. Instead, they consist of periodic oscillations of electrical and magnetic fields generated by charged particles, and can therefore travel through a vacuum. These types of waves vary in wavelength, and include radio waves, microwaves, infrared radiation, visible light, ultraviolet radiation, X-rays, and gamma rays." Wikipedia.

I'd compare wishes to the electromagnetic waves.

They hypnotise us easily. Therefore, I think they have electromagnetic properties. As the sound waves, our wishes, as well as every thought, propagate the media. The electromagnetism of our thoughts, of our self physically is of a tiny dimension but is exist everywhere.

In my childhood, I remember, young ladies from our neighbourhood, my sisters, cousins used to gather at our home and play the divination games. They were all young student girls and they were very curious about their future, how would their fiancés look like, what letter their name starts with, (in my country marriage usually arranged by the parents or family members). How many children girls would have, what job their future husband would do etc. That was fun, girls were pretending to study at my Dad's library, pretending they are going to read for exam preparation, but would often tell each other's fortunes.

They drew the diagram in form of a circle with the letters in alphabetic order around the circle on the large paper sheet. Then each girl would write her question on a small piece of paper and pass it to the dooms teller (usually my sister, Nargiz, the incredible actress from the birth). She read the question, and then held the threaded needle on a thread about 30 cm long, at proximity of 2-3 cm above the diagram.

The needle, like magnetized would move from the letter-to-letter, around the diagram, and spinning around certain letters. Those letters the girls would write on a different paper and the order of the letters, the needle choose usually were compounding comprehensive words like there was someone talking to the group of crazy girls, who genuinely believed they are talking to their relatives who passed away.

Yes, the detail I have almost forgot is that they used to address the questions to the named celebrities or family members who passed away. That is what our culture and religion bans: no one allowed to disturb those who passed away unless to wish them peaceful rest. I can't remember if any of the foolish questions the girls asked and the answers they have received actually happened, neither I am able to confirm the outcome of the fortune telling, even though they did ask questions about me, but what was always wandering for me is that the speed of the moving needle that would increase with the number of attendees of such evening quizzes.

Without trust in any word the girls did received, as the same questions would be answered differently each time they were asking, I tried to hold the needle myself many time. Without moving my hand, each time the girls asked question, the needle would head the thread to the different letters and clearly spin around certain letters on the diagram. The clear feeling I have had at that time was that the needle followed kind of magnetism. I did checked that many times: my naughty sister did not place any magnet beneath the table!

I think that was magnetism of girls' wishes. Even though the magnetism of their wish waves was tiny, it was able to move the needle.

To my logic, it may be true that wishes, being physical tiny sound waves, own property of magnetism, too.

You can conduct the same experiment with your friends: the threaded needle moves as that is within the magnetic field if surrounded by group of individuals.

At some point the thoughts that fall in our acceptance, may transform to the wishes.

Replaying, reiterating, remembering the thoughts or scenarios makes us accept those scenarios and bring them into our lives even if we are not aware that by replaying we turn our thoughts to wishes.

Whose wishes are they: ours or are they of Our Imagination, which is the Supreme Reality, -remains an open question. But there is no doubt; established mechanism of creation of things and events is through thoughts and their repetition.

By repetition we virtually can transform any though to the wish. Provided that the wish has electromagnetism, it would work towards the physical exhibition. That is how we attract events in our reality.

Our ancestors have already had the perfect understanding of such mechanisms and even postulated the process in religions. Therefore, the pray and basic principles in all religions established: these are very powerful protection against our own selves.

By praying, by asking the God to forgive our own (bad) thoughts, we are protecting our selves. By wishing every good thing to happen to our neighbour, we bringing good events to our lives. By giving, we receive. And yet in every religion you will find the main principle: do not harm.

In Persian, "God" sounds "Hudo"- The Self. Our own imagination is much more extended then we ever able to comprehend. In that extension, within which we all exist, there are lots of vibrations, created by our Imagination. And only we are, our Self, who creates borders for our Selves or we are, our Self, who opens the Pandora box.

In today's society, unfortunately, we are entertained by the movies that almost glorify the murder, the crimes and abuse. In decent society, such entertainment shall denied by the public itself rather by any censoring body, because people would know the precise mechanisms of the creation of events in our lives where they will not allow any thought of any harm to disturb the harmonious process-happiness.

If faced with the victims of crime, let's help them, wish them very best, give them anything you can, don't give your energy to the crime perpetrators. Do not allow their vibrations into your life.

Our wishes that come from sincere feelings of love, forgiveness, and joy are the mighty protectors and builders of our own better destiny. They create powerful energetic corridors, where, once we get in, we are safe.

Once accumulated, vibrations of positive wishes and thoughts transformed in energetic supersonic tunnels of affluence, joy, happiness. Once you've realized their existence you will keep your way within those corridors.

It is a model of the God's feelings and thoughts – to wish everyone happiness, joy, pleasure. God want everyone to be happy, so that is why whatever our intentions are, they become true, particularly if replayed in our minds.

At the same time the God's Intelligence there is no doubt is Supreme and all knowing. But the way our intentions come into our lives are simple ways of basic physics. Should we wish something bad to happen to our opponents, we will be the first to suffer from our wishes, should we ask for their prosperity, we are the first to encounter the affluence. By wishing the very happiness to our enemies we become happy first.

Our ancestors new that simple rules as the rule of thumb and were praying for others good in Churches, Temples and Mosques. They knew there are the cords in our brains that play constructive or destructive music and they knew, how to manage the rendition to benefit not only themselves, but also their society for their own good.

How did they know we all are The One?

Those positive wills we wish for someone else, those sound waves create the powerful vibrations that act as a whirlpool and engulf in the affluence the wisher in first. It happens if we wish those good things to happen for ourselves.

Let think what happens when we wish to improve, or when we process the thoughts of our shortcomings, that we believe are our defects, but really only God knows if those are our faults or our talents.

By replaying thoughts of worthlessness, feeling guilt and shame we also create whirlpool of negativity. It works same as the thinking negatively about the others. Such thinking indeed creates whirlpools of negative vibrations and hence bad events.

We encounter greater problems with our health, with our jobs, finance, and come across the bad luck and so on.

In negative (bad) vibrations, our actions stipulated by the scattered brain activity and most of our actions are erratic, wrong. Let's repeat it again: what turns the thoughts into the action? A wish, an intention. I think those are enhanced vibrations, electro-magnetized by repetitions.

Thoughts, whether created by us or existing independently, create vibrations, then turn to wishes. Thoughts we replay in our minds create vibrations that reflected by our brain cells and activate parts of our DNA.

By creating positive thoughts and wishes we become the God's Essence – The Love. Whatever you call this Essence, all and any of the most beautiful words describe The God.

Even without sincere wish to end their life, depressed people somehow fall into those vibrations, where repeated thoughts turn into intention to harm themselves and then a tragedy occur. Vibrations act as a whirlpool. Once trapped in negative (bad) vibrations, the individuals might not be able to get out from them without professional help. Other individuals, who fall in such bad vibrations, without depression being prominently manifest, may face the range of different diseases, degenerative changes, financial difficulties, and even heavier losses. I think these are the traps we fall once caught in negative, bad vibrations.

Even if anyone from one family has fallen to such vibrations, through the DNA, those bad vibrations may affect other family members. It is because the DNA's structure, while being known in reasonable extent to the science, remains such a mystery and literally resembles to the written words. It contains loads of genes, which in different stages of our lives will act in different scenarios.

For example, at some point of our lives the risk of activation of the cancer genes will be higher, while during other stage the gene responsible for achievements in sport will light up and continue to keep active for many years ahead. That will depend on the vibrations; we are under the influence of, at those stages.

In other words, that will depend on your choice through which corridor you want to pass your journey: the tunnel of affluence and love, or the corridors of negativity. Once you clearly realize how your thoughts selections navigate you to the different poles, you'll get through with confidence and choose the right way.

The most correct compass is your feelings. If you feel you are happy in place, you have chosen, or about to choose, then you approaching the right tunnel, right vibrations corridor, right quantum location to leap.

Just think about this.

Talented children might not be able to exhibit their talents if caught in vibrations of despair, while the others, raised in happy environment, full of hope, will fully develop their talents.

This doesn't correlate with the social status. Look at the biography of famous footballers, artists: are they all grown in wealthy families? They have found their own pathways of happiness to be at right place, the right vibrations that allowed them to explore their talents.

Our thoughts, intends and words create all vibrations we are in. Our words, and deeds – all are almost equal to our thoughts and intentions in creating those vibrations, we are in.

Those vibrations firstly affect our bodies and lives, then lives of those who carries our DNA – our family.

I'd rather refer the vibrations we are under the influence of at every moment of our lives to the circumstances.

We are living in vibrations, even though their measures less then material. But we are living within the, and under the constant vibrations. I would say, the whole Universe is vibration.

Every thought emitted travels as the tiny sound wave and vibrates the tiny particles of our cells bricks-the quarks.

The choice we make is whether to fall or not under their influence and we can choose on which vibrations / frequencies to switch to. By choosing right words, right thoughts we make our choices to enter into the influence of good or bad vibrations. We can make our decision whether to fall into the vibrations of health, wealth and affluence or to the vibrations that can lead us to the poverty, despair and disease.

However, even if we catch the disease or face the serious illness, the choice is ours to switch to the vibrations of healing and happiness.

There are solutions for everyone. Speak to your doctor, contact the yoga instructors, meditate, and see your friends. Seek positive vibrations, seek happiness, create your own vibrations of joy, and find out what makes you happy.

Emit vibrations of happiness and hope. Explore the bless of existence. You are here to emit vibrations of the happiness, joy, hope.

When we dream, we create, we vibrate in excitement. If we don't dream, we stagnate our creativity, the power given to us as a necessary tool for the natural evolution. The power given by the God, by The Mind, tremendous power, even though it's invisible ...

The burden of this power turns to the low mood and depression, if unused. So whom do you think blessed more: the individuals who live in aspirations or those who live in enormous wealth? Who would use their dreaming potential: those in need or those who have everything they might want?

Do all people, living in wealth, use their creative power in full?

Where in the world, do you think, vibrations of hope or those of the desperation should be intensively created? Does that depend on the financial stipulations, the politics or on spiritual beliefs?

5.

Bionics

Bionics is the science, which uses biological methods, found in the nature, to apply for use in technology. Thanks to the bionics, people explored how blind bat swifts its way speedily before the obstacles that rise in it's way.

That was a little blind bat that had inspired humankind to use the sound waves it emits to recognize the obstacles on its way and move safely without using its sight. Thanks to that discovery, the ultrasound diagnostic technique was developed and used until nowadays.

Bats are blind but they fly very fast. Little creature emits sound wave which is reflected with different intensity by the objects on it's way and transmitted by the air back to the bat's ears indicating to it if the traffic is free to fly.

If any obstacle arises, the wave, reflected by the object, comes back to the bat and it's changes it's rout to avoid the clash.

Think, the bat navigates with the sound in space, doesn't that remind you how we use our intuition, the sense of comfort and discomfort before making an important choice?

The bat avoids the areas it feels by the sound waves it creates. So do we. We create the thoughts, and we feel, our mind feels, where is the space, or domain, or let's call it a radio frequency to tune to, which is right to our thoughts ... And there we direct ourselves, or tune our minds to those frequencies, like a radio antenna there our mind directs us. Are we always aware of that?

The other great thing bionics has implemented for us, is the Doppler ultrasound. Scientists learnt that stars moving in one direction in the skies reflect the light in the color, which differ from the color reflected by the stars moving to the other direction.

Same principle applied in use of the Doppler scan to locate the blockage in the blood vessels, because the direction of blood flow will colored in red and blue, pointing exactly where is a thrombi formation happened, or to detect where the blood vessels communicate abnormally, or where in the heart is the problem blood flow may indicate.

The scientists developed the use of the ultrasound further.

If focused, sound beam can destroy the stones in kidneys without harming the organ. More focused, it incinerates "clinical waste" and other hazardous substances.

Bionics for me is a field, where the ideas are coming from, the methods are explored, but they have already existed there, in the nature.

That is a matter of a right time and circumstances for us, for the humankind, to pick those methods and use as a solutions for the problems, we encounter.

Think about it: the idea of ways on how to deal with the problem was always existing in the nature, but only with the technology advance, with the creation of the Science, with it's branch of engineering, later called bionics, that become feasible to notice and apply those methods.

In fact, there are many things are feasible to apply for our lives if we are in a right environment.

So ultrasound waves not felt by our receptors, non of our receptors able to pick up, to sense those tiny wave signals, but our organs receive, transmit and reflect those waves. As a result, we can see and assess the images of the organs seen on the computer screen!

This makes me assume that every organ tissue reflects even tiny waves. Organic tissue (that made up from the live cells) also transmits, I also believe, processes and fall into the influence of the waves, emitted by other objects. But there have to be also a sort of cells in our brain, which receive thoughts (the sound waves), and process them, even if we are not aware of that. It is my belief that the sense of déjà vu comes from those preceptors of the thoughts waves cells.

They must be similar to the "inner GPS cells", which are responsible for our spatial orientation, or our location in the environment.

Cognitive map cells were discovered by Professor John O'Keefe (London, UCL) and situated in the hippocampus, the part of the brain, which is also responsible for short – term and long- term memory.

Professor John O'Keefe awarded the Nobel Prize in Physiology or Medicine for his discovery.

'GPS cells' responsible for animal's behavior "on the basis of distances and directions towards desired goals or away from undesirable objects and the locations" – his researches revealed.

*.(UCL http://www.ucl.ac.uk/cdb/news/cdb-news/news)

Scientists nowadays develop spare parts of the body and organs for those who in need for them, all thanks to bionics!

To me, if I say that in simple words, the bionic is the science to learn from the nature, to notice the choices, which are already, exist. The Supreme Intelligence gives them to us; King Solomon had already devoted his book to it, to the Wisdom.

17   For he hath given me certain knowledge of the things that are, namely, to know how the world was made, and the operation of the elements:

18   The beginning, ending, and midst of the times: the alterations of the turning of the sun, and the change of seasons:

19   The circuits of years, and the positions of stars:

20   The natures of living creatures, and the furies of wild beasts: the violence of winds, and the reasonings of men: the diversities of plants and the virtues of roots:

21   And all such things as are either secret or manifest, them I know.

22 <u>For wisdom, which is the worker of all things, taught me: for in her is an understanding spirit holy, one only, manifold, subtil, lively, clear, undefiled, plain, not subject to hurt, loving the thing that is good quick, which cannot be letted, ready to do good,</u>

23 <u>Kind to man, steadfast, sure, free from care, having all power, overseeing all things, and going through all understanding, pure, and most subtil, spirits.</u>

Wisdom of Solomon, Chapter 7

King Solomon refers to biology (circuits of years in the trees), zoology, astronomy and physics, - all arranged with the ultimate wisdom and purity.

We have to look around for the solutions of our problems, they already exist in the nature- that what the bionics does.

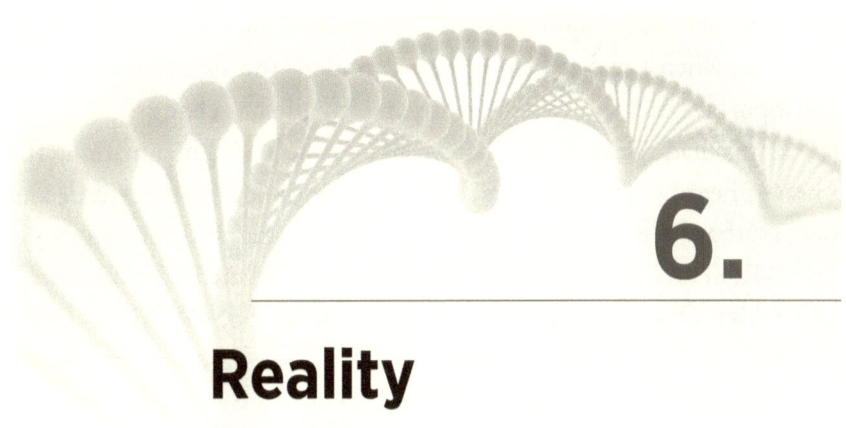

6.

Reality

My understanding of how the reality exists came long before I 've started to learn more about the quantum physics.

I remember myself waking up in early morning in November 1ˢᵗ, 2003. I wanted to write down the words that I've seen and clearly herd them in my dream. I wrote down that dream and still have that note.

"ARE YOU (YOU, THEM, EVERYONE) REALLY THOUGHT YOU ARE OUTSIDE OF MY CONSCIOUSNESS/AWARENESS?"

That was a whisper, or though or something that much SO NATURAL, that I can't even describe, how subtle that voice was.

I had complex relationships in my life by that time and found myself in a difficult situation. It was my time to grow out through the obstacles, which I was simply unable to recognise as was never faced them before … also no one told me that through the obstacles the individual evolves. But that is a natural process of our development- facing the difficulties.

I was asking myself: why things are happening not in a way we want them to be, why you'd have to sacrifice something which you'd think by that time is the best for you, but many years later I'd realize they are not …

My relationship was not a best thing to happen for me, my decisions I'd make sake for my family and would think I'm sacrificing was the best and the most right thing I'd ever done in my life. That very moment, when I realized things in my understanding are wrong I saw an important message in my dream and my life has changed.

It was made clear for me that we all exist in God's Imagination. The phrase I heard in my sleep was so sharp, so sensible, and that subtle and natural that I felt had to write it in my diary at the next very morning.

"ARE YOU (YOU, THEM, EVERYONE) REALLY THOUGHT YOU ARE OUTSIDE OF MY CONSCIOUSNESS/AWARENESS?"

Ah, the God is the Greatest mischief, They have made us up in Their imagination and hence we all exist there! –I was joking. But God is The most Beautiful, The Most Wise mischief, The Most Wonderful, who always knows Everything. Because God is The Knowledge, is The Imagination, is The Choice, is Everything, is Everyone.

We all exist in The God's imagination simply because God allows us to be within God's Mind. Since we are within, we too, carrying power of imagination and hence creation. Although we create in our own way, sometimes sporadically, sometimes erratically, sometimes in unawareness, but we create in a ways we assume are better for us. Those assumptions often shaped under the influence of our environment but not necessary always work for our goodness.

Actually, later, when I heard that Sir Albert Einstein had already told out this hypothesis- we all exist on God's imagination, I was glad to know that.

God bless you, dear Einstein, for the discoveries in physics, understanding the truth as it is, and for your Kind desire to teach that to us!

So, what I was thinking about the world is true! I also believe that thoughts and ideas travel in the air, space, time and all other dimensions.

Let me explain my understanding of fundamental basics of our existence and what I call an immeasurable reality, measurable reality and the mirror reality.

That what I revealed for myself, we all exist in God's imagination, which is infinite indeed and has that power everybody know. Whatever you call it, by the mean of the all the beautiful words, you will be actually right.

"Whatever People think about Us – that We will Be"– That what the God tells about themselves in the Quran, in Torah, in Bible and in every religion.

I call that God's Mind an immeasurable reality, because it is non-material. But it's more real, than our, physical world. Every quark, every thought, every intention remain in that reality in it's pure, original condition.

The modern quantum physics supports the theory and calls that reality a Quantum Field.

Quantum physics state that all atoms consist from the quarks, that only appear when we observe them, and they are non- local. They appear and disappear at the same time and their position is only a possibility. So that is to mean that even the quark is located anywhere in the space, its position is only a theoretical possibility.

Having accepting that the thoughts are the sound waves that able to vibrate the particles of the media (not the molecules, but the quarks) the probability of the quarks position in the reality would be stipulated by our attention and hence, by our thoughts. As by navigation, by our thoughts we move towards the vibration and create our vibration fields around us. The fields created by the vibrations, I'd stipulate as the negative and positive, or, in other words, destructive and constructive.

We all consist from organic cells made up from the molecules. Molecules are the consequence of the atoms. Atoms consist of the elementary particles, the quarks that constantly move in a wave fashion.

The continuous nature of our waves makes our images not flicker between the measurable world and the immeasurable One, within which we all exist and coming from (that is a God's imagination) and reflection reality.

We are connected directly to the Supreme Imagination, to The Supreme Mind, by mere act of thinking. Since we usually always think, even when we dream, we are constantly connected with the God- with the Supreme Mind. When we aware of our thoughts, we are in a fine linear reality where we make our choices, then we approach a Mirror Reality. We are not flickering since waves are continuous and reflected by the mirror reality. If the Science proves that every single quark, every single cell has own Intelligence that connected to the Supreme Intelligence, then we are a very perplex creatures yet still we all are The One. And then, logically, the Mirror Reality would extend to the every level of our existence –from each and every quark to the Universe.

So we are continuously present in measurable world thanks to the reflection property of the waves, too.

The quarks- the elementary particles, everything is consist from, are constantly moving in waves fashion from the immeasurable reality to our space and time reality (or reality, restricted by the space and the time). I would refer the space and time reality to the infinite, physical reality; but again, according to Einstein's general theory of relativity, the reality which is the source of everything (Supreme Imagination) and not complies with the space- time – gravity dimensions as it superior to the material Universe, is more real. We are yet to embrace that by our minds and to accept that.

That is my understanding of what nowadays quantum physicists call "the principle of non- locality", the aspects of those principles. But while the particles are moving, constantly appearing and disappearing in this world, the same very fundamental principles shall apply to every material entity: at the level of a molecule, the cell, the organism or the whole city or a bit of information, like a thought, which is a wave in my view.

Then, while consisting from the tiny invisible quarks, we are, too, constantly moving in Conscience into the Imagination reality and to the physical reality again. But we exist in both worlds at the same time.

It takes no time for us to move from imagination to existence in the measurable reality, **because in Immeasurable Reality the time matter does not exist.**

The line between us and the Infinite Conscience (that is the God's imagination) is like a mirror. In my understanding, that is a line, where exchange of the energies occurs. At that line, where thoughts meet physical reality (perhaps, God's thoughts), from there light glows. There biophotons can be measured, at physical level. Logically, electromagnetism with its applied properties shall be found here.

There also lies the line, where we have made our choices.

It is where we make our choice, where we judge, where we make our decisions every second and our choices come back to us at some points of time.

That line exist at each level of our world, be that a quark, the molecule, the cell, the whole organism, the Universe.

So, for the God, there is no difference between the single cell and the whole Universe – that much the powerful is the real reality.

For that reality, whatever we call it, The God, The quantum filed, The imagination, in which every of our cells, every single cell, and everyone of Us is an equal contributor, Time is also not of any importance, it simply doesn't exist. It is eternal, and endless.

But for the product of that imagination – current physical world we perceive as an only single reality, time is a matter.

We all coming here because we wanted that. We all doing our choices every second without knowing sometimes the very purpose of our journey, as we usually detracted by things, circumstances, wrong emotions and wrong thinking. This happens because we don't posses reasonable knowledge and we do not pay attention to who we really are. I believe in correctly switched attitude that would be easier to develop our own ideas of who we are and find out our unique abilities to do our mission, to do the job we are born for.

It is not necessary to be related to our profession, but that might be the role in our family, community, the social circle.

The correct attitude, in my mind is the harmony of wishing only the best to every of our own cells, to any person around us, to the whole Universe.

Modern Science moves towards the shift in the philosophic, theoretical and practical understanding of the fundamental principals of our existence and our interference with the environment.

We are getting deeper understanding that subtle forces and their interaction in fact, much more powerful then more prominent exhibition of those forces. I'm talking about creation of our world, about it's evolution and our future.

Here it's worth to mention again that we make our choices every second, because our mind never stops thinking.

We locate the quarks of our self with our thoughts and at the same time, we are allocated in the frequencies, (vibrations), of affluence or losses, of joy or sorrow by thoughts. Thoughts that fly in space-time, that emitted by us, by those who surround us, create whirlpools of Reality, prosperous or devastating. Whatever you call them: the whirlpools, the tunnels, the zones, they have their own electromagnetism. They attract and engulf in their space, caused by the exchange of the energy. The repetition of the thoughts since the thoughts are the sound waves and they do incur their energy moves us towards those tunnels.

There is a stunning consequence of events of individuals who falls in the certain vibrations. I compare such fields to the whirlpool. It often happens to us: who feels low, worthless, who pays too much attention to analyzing past events feeling sadness, those more often face financial difficulties, some heavy losses, physical disease to name a few. The others, who are in a positive thinking wave, on a positive vibrations, overcome disease, financial difficulties, they progress to success in every area.

So when we make a choice, whether we pray or meditate, we are likely to receive the best intuitive choice option if we do not violate certain principles, like harming others or harming ourselves in any mean.

Therefore, wishing the very best for others is the vital point as a base on which we are to make our choices. We don't have to wish everything best for everyone around during our praying or meditations: take it higher; we are to develop that attitude as our normal state of mind. For me it took ages, for you it might take as little as one months, most of us have already born with that attitude, some others been received that from their families.

Simply start your morning from wishing every beautiful thing to happen to your kids, to your husband, to your Mum, to yourself. Once awaken, lie at your bed and take a few seconds for daydreaming.

Think about best things you'd wish to those whom you love. Write them, if you want.

The iwishyou all the best things to happen is a very intrinsic to the Christianity but again, everyone can develop this virtue as a main pathway for the personal success and to achieve personal aspirations.

I want to mention that chasing our goals shall not to turn to the sort of obsession. You wish the outcome of your goals, you wish the state of happiness. But we don't always know what is the happiness for us at the moment. For example, you think you have to buy a huge home for your happiness. You need to earn millions in order to achieve your goal. Think what exact happiness would be for you in that huge home: full of friends, or kids, or pets, or people around you who work for you? You may end up in a smaller house with the batch of happy kids but the main point you don't know what is better for you, you have to experience that in order to know.

So, I suggest wishing true feeling of happiness then dreaming about the millions in bank obsessively.

The state, in which we make our decisions, is a very important clue to our current situation. If we are in state of anxiety, low mood or even in depression, the choices we make for ourselves and for others are far from perfection. That how we come across the situations and people we have in our lives. That might be a single gesture, or face expression, which the person opposite has judged you upon, or it might be an issue regarding your credit loan – whatever is the situation, we always make our choices daily, every second, even when we dream.

We find ourselves unhappy, unsatisfied, hurt due to the miscellaneous reasons.

Depression is the main curse of modern society and it's elements lie at the bottom of pathogenesis of every disease. To come out from low mood, from depression, to attain perfect health, lasting youth and anything else you might want you need to know the basics of our existence and interfering with the environment.

When we know how things function in our world, we can influence our life and body states by using natural effects, which work irrespective of our opinion about them. We can do so by knowing, remembering and reminding ourselves the physics of our thoughts.

Whether you want it or not, accept that or deny, where ever you live, in South Africa or at North Pole, you are under the influence of natural laws that rule the whole Universe.

Knowing their basics will help you to navigate towards your goals. Knowing how the Universe functions will make you feel grateful for the things in your life as they are, once you've become aware of them and start use the nature law in your life.

We are nowadays approaching a brand new world, where the subtle forces proven to manage the Universe and while invisible to the naked eye, they do exist, they are a fundamental basement of our existence and they do manage the Universe functioning in it's way.

The thought, provided that is the sound wave and complies with the physics of the sound wave, propagate through our mind, vibrates it in a certain manner, vibrates all of our organs, blood, the cells and passes our DNA. How every thought affects our body? Every thought would vibrate our cells and the molecules of our DNA, oscillating gene's particles, swinging the quarks in their atoms… In very subtle manner, thousands times lower then the lightest whisper, but no doubt, it would affect our mind matter at first.

So, think how is important for our brain to wish everyone good things to happen and how that may affect our health- disease state of mind.

As the sportsperson flexes muscles in order to keep them fit in shape, everyday we have to process only positive thoughts through our minds. That how we develop constant attitude of our emotional and intellectual states and also how we tune our brain's antennae to the constructive vibrations of affluence.

If things are not going our way, it's better to switch the mind to the dream, where things are exact that way we want them to be. In your imagination, which is your private laboratory of your happiness, make the images of how, for your happiness sake, things should be. But remember to not harm anyone in your personal laboratory. Don't take there people who made you unhappy, but don't wish them anything bad too, let them go.

Don't forget about the physical exercises.

Physical activity is proven to boost healthy chemicals exchange in our bodies and also one of the powerful mechanisms to heal almost every disease.

In disease state, please, do not abandon your doctor's advice. Take every available opportunity from conventional medicine. You will meet people who wants your healing. You will use the medicines that proven to help. Take every chance and use it.

What kind of future we face?

Nowadays, when robots are performing difficult surgeries in some hospitals, when anyone can buy any stuff worldwide by clicking the touchscreen button, where news spread out by social networks in seconds, what kind of future we can desire to approach? And what it would be?

In the future, cloning will be a subject taught in the classroom by practical lessons, wealthy elite would own planets, universities rent those planets to teach their curriculum to the students, time dimension would become a stipulated component and yes, there will be no death, no disease and every person affected by any kind of disorder (like depression, as I believe, depression is the only disorder the humanity may take with itself to the future) will be treated at the national emergency level and will get 100 % cure. There will be no place for hypocrisy and political games in that world. Every politician will have to be a genuine hero to get more votes, and will work in scientific laboratory, unlike it happens now, where wealthy elite creates the villains and fights them in order to appear as heroes. There will be no place for lies and prejudice.

The heaven is right there. And we all will witness it.

There is no doubt we all are moving towards that. Because that is a natural process of humans' evolution.

Is our Universe is perfect? Are our lives perfect? Is your health ideal? What about the health of your beloved ones? Would you like to change your life?

Everyone wants to live full life, most of us dream of endless wealth, better position in our career, to remain young forever. But what about those who possess enormous material wealth? Is their life is ideal?

Of course you know the answers but if you wipe off the prejudices and past experience from your mind and try the simple techniques by reminding you the knowledge I'm offering in this book, the new world will become widely open to you. Very Soon.

7.

Emotions

"The finest emotion of which we are capable is the mystic emotion. Herein lies the germ of all art and all true science. Anyone to whom this feeling is alien, who is no longer capable of wonderment and lives in a state of fear is a dead man. To know that what is impenetrable for us really exists and manifests itself as the highest wisdom and the most radiant beauty, whose gross forms alone are intelligible to our poor faculties - this knowledge, this feeling ... that is the core of the true religious sentiment. In this sense, and in this sense alone, I rank myself among profoundly religious men."

Albert Einstein

Emotions are triggered by certain thoughts, by circumstances and people around us. If we replay thought continuously, we embed them into our DNA by turning on the genes that correlate with the certain words, thoughts and emotions attached to such thought.

In return, that would be our continuous /permanent state of mind. What do you want to be emotionally and mentally? Replay those thoughts which make you feel accordingly, regularly and that you will be.

Emotions also may embed in our mind through our childhood, and we live in certain emotional status as we get used to it.

But here it is our responsibility to decide: is the emotion I'm currently live in, my natural state? What I want to be like? How can I achieve true happiness? In other words, what is the happiness for me?

It doesn't really matter; in what age you ask those questions. Every individual, I believe, born with the knowledge. We know when we born, who we are and what we want to be. But with age, with experience, or under the influence of surrounding people, we replace that knowledge with the range of other theories.

Some children do not take other's opinions and they know exactly what they want to be, what their happiness would look like and often such individuals achieve their goals.

If we are lucky, our parents will navigate us correctly using their wisdom to the right vibrations of developing our talents, if we are not, we can navigate there ourselves, at any age, provided we know our potentials, and we know how vibrations meridians work. Wishing every good thing to everyone around and to ourselves may help us in navigation to the meridians of constructivism and prosperity. The right emotions then would follow.

Surrounding children with the sincere love will make the child flourish, because in true happiness state only the child with no any hesitation will find the right way to discover and implement his or her talents.

There are plenty choices of the thoughts you create and those you think you create. Beware, if the states you want to be, harm others, your hippocampus will start it's poisonous work, as the intention to hurt will be reflected to your body, your brain, your DNA.

If your intention is to thrive, then sincerely wish to thrive to others. It might be difficult to wish to thrive to the tyrants or whom you see as your opponents, but the cultivating such wishes is the pathway to happiness, to joy, to the evolution, to the right frequencies, where everything is possible.

So what happens when we don't want to see anyone, want others to leave us alone? That must be our defending reaction, when we need some space; we turn to the lower mood.

Our brain needs rest in order for our personality to grow. At that time I'd advise you to read. Reading is a perfect way to evolve.

Emotional abuse of others is a crime, I do agree but everyone can fight it and win that fight rather then get into the whirlpool of negativity, depression and all other consequences that entails.

Envy. That is a destructive negative emotion, if misused. That arises from the thoughts of someone's superiority over you. But since that quite natural, let think where it comes from and how can we use it for our own growth. It was an envy that showed us in wild past what we need.

According to the theory of Charles Darwin, evolution was a slow process, a ladder, which brought the species to their current conditions from lower forms of existence to the more improved ones. In the Quran (Chapter 4:1, Chapter 24:45, for example) it was repeatedly mentioned, that God has created all animals from the single cell, and that all organisms originated from water.

I usually tell to my friends about how to coup with the jealousy, envy the following story: that was a lizard that spotted a flying insect creature and envy has sparkled an idea for the lizard to develop own wings. That lizard grown own wings and learnt to fly, and turned to the bird that slowly became a swan ... Doesn't Charles Darwin theory tells us approximately the same story?

Envy to me is a signal that I'm unhappy in some aspects of my life. I need to seek what makes me unhappy? What triggered my envy? What can I do to get same property/ appearance/ relation / whatever brought my envy in existence? How can I improve to achieve similar subject/ situation/status?

Envy to evolve further is a healthy choice but envy in order to hurt hurts us.

Only thinking of how I'd feel if I had same thing that triggered my envy in my mind eases my disturbance. But I don' have to stop there. I want that to be in my life even better. What shall I do for that?

Leave your thoughts of harm and destruction, but create instead, and you will get better possession /relation/whatever as you saw the first only variety, but next is always better!

Envy about someone's appearance? That's easy to deal with! My Dad used to say that by physical exercise one can achieve EVERYTHING in health : body appearance, rehabilitation from any, any diseases and so on.

I was overweight in teen ages but since started my exercises and importantly, kept continue them for the years, each time while exercise, was dreaming and literally programing my body to appear at my 40s as I'm 20 years old. I'm 40 now, mother of three, happy with my figure and I programming my body appearance to be as I'm 25 in my 60s. Still have 15 years to go and to achieve that! And still exercising regularly.

It is important to not over-exercise at the beginning of your training. The body must not remember the exercise as stress, but as a pleasure. Otherwise, the next exercise would be not welcomed by your body.

I think that every programming (like meditation) during physical activity or in complete silence and relaxation will work very effectively. It will also balance emotional state.

Physical activity takes our attention away from the stress, as every meditation technique does.

Some of my patients who visit me on a regular basis for the prophylaxis check-ups have been diagnosed with forms of autism before. They naturally don't pay significant, or any, attention to the stress. They are very successive in their professional field and their wealth makes them enjoy life fully with their families. People who can acquire such ability to not pay significant attention to the factors that makes them feel sad, some sort of egoists (in a good meaning) are very successful in their achievements. They help to those afflicted with the stress, but they do not go deep into the empathy. At the same time, they do not judge, having not crossed that fine line of indifference, they have the true sense of helping to others and to any who asks, but they are far from the judgments.

That what I learnt from my affluent, talented patients: they help others, with what they can, they wish others very best but they do not empathize too far.

But what I suggest, you pay attention and feel very good things to everyone you encounter in your life and keep positive.

Most of us dream about feeling carefree. My message is that the carefreeness is achievable with pure desire of all good things to happen to others.

So isn't that a proper, superior science, the essence of which we have to learn through our lives and teach that to our children? Isn't that what all religions teaching us? The initial purpose of every religion is to teach us to think in right way and to fear our own thoughts, when they are negative. Indeed, every religion teaches us to live in a peace and love to the neighbor. The main purpose of every religion is to not harm in any way: "That who had killed one innocent soul equal to that who killed all souls over the world"–The Quran, Chapter 5: 32.

Lets tell again a word about depression. In my view, depression is a burden of the great power- the dreaming power, the creativity, which is unused and hence stagnated. Depression also can be as a result of severe loss, a tragic event, bereaving.

But there are natural mechanisms that heal our souls and if they don't work in six months time, you need to seek professional help.

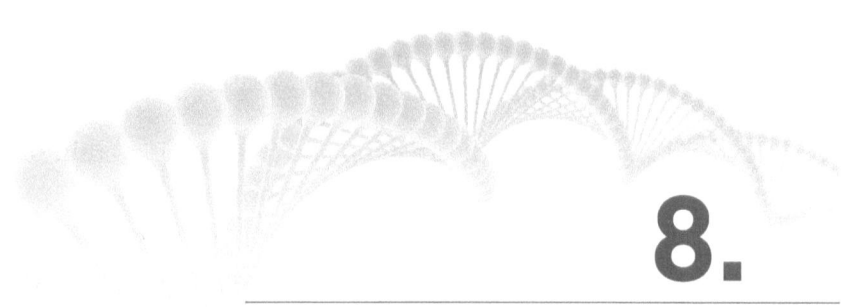

8.

The Instincts

Humankind does make a choice almost every moment of their life. Even though not always aware of it. We make our choices according to our intuition, when we aware and to our instincts, when we are unaware.

It's true to say that our thoughts are the waves, as the quarks are waves, too. The quarks are not fixed rigidly to their positions in space, but they flicker under the microscope. Their position in space (just think about the statement) determined by our attention.

When someone looks at the quark, its position can be located in space. I remember the article published in 1980s describing the experiment the scientists conducted to locate the "electrons", as those were called at that time. That was an article in soviet newspaper and seems to me now, as the Soviet Government, closed to the whole word at times, was ready to prepare its nation to the groundbreaking discoveries in the Science.

The experiment held in America. Scientists placed the tiny quark between two fixed cameras and tried to photograph its position in a space. When human was watching to the camera, the position of the electron was captured in one location, but when the snapshot made by the unmanned camera, the electron was captured as a wave trace of it's probable position.

The experiment was indeed repeated hundred times and each time the data shown the same result: electron captured while someone was watching at it, had certain location and it's location was unclear when no one was watching to it.

If discussed further, I shall tell here that their position determined by the mental intelligence of The One and The One is WE ARE- the post that quite complex to comprehend. But we don't need to dive deep into that level of the science that can make us confused.

That is to mean that by our attention, by our thoughts we are literally move the location of the quarks (from which our cells consist) of our bodies, making the cells to work under the certain conditions. It is difficult to define in what time such movements occur, by virtue of two possible answers: as that happens every moment and as the time matter simply doesn't exist in non- material, Imagination Reality. The changes or our cells may hence occur within several years or in no time. Here is one additional comment: the time repeats its circles as the seasons of the year do.

And yet that is to mean we are constantly moving towards powerful whirlpools of the tunnels that have always been existing in the Universe and beyond: the frequencies of joy, fear, carefreeness, despair. At some points of our life we jump from one tunnel to another, and hence move the events in our lives. At some other moments we are hanging out at those tunnels, we discover them for ourselves.

And one day our discoveries allow us to comprehend, to accept and to embrace the reality. We all are The One. We are here to make our mission- perhaps, to contribute into a harmony of the perfect reality. In this reality, only our best wishes are true and counted. No deed, no thought, no intention vain away. So we have to carefully choose the zones of feelings, our thoughts bring us to in order to physically be better.

It might sound strange but if Albert Einstein was right, we all are come out from the imagination of the God, hence we are all The One.

While we exchange out thoughts waves with the Supreme Intelligence, which has created us, by mere thinking, I believe, there exchange of energy occurs. Replayed, thoughts acquire electromagnetism property and thus the tunnels/zones/meridians of thoughts frequencies would possess their electromagnetic power, too.

The nature of waves implies that the length of the waves is inversely proportional to the frequency and that the propagation velocity depends to the density of the tissue / environment density. So if the Immeasurable Reality doesn't have any density, it takes nothing for the thought to propagate there through to the measurable world? At the same time, our thoughts, as the sound wave, travel in our physical reality, and permeate physical tissues, interfere with the other waves and even create whirlpools of the affluent of destructive zones.

Otherwise, why we feel the "positive or negative" atmosphere in families, person's mood, in the towns and elsewhere? That is not what we judge through the facial expressions and body language, but, perhaps, is an intuition, what we feel, rather then the visual and audial perceptions that we judge on.

Thought waves own same properties, those universally apply to all sorts of the sound waves. The lower the frequency of the wave, the longer it's way and more stable is our brain, more calm is our emotional state, more productive our work, more far in attention we go.

Thoughts of analyzing are shorter then the dreams, because the dreams spread out to our lives throughout our childhood, but thoughts impact us in the moment we experience them. The longer the wave – the lower the frequency. By dreaming and by wishing, you make your brain (sound) waves lower, and waves become coherent, too.

The wishes and intentions, I believe, are also reflected by our photons – tiny light particles we emit, that compose what some call our "light body".

Although the "light body" term refers to the spirituality, modern physicists refer that to the physical entity. According to Albert Einstein, photons are elementary particles, that emitted by every physical object and interacts electromagnetically and can be observed micro and macroscopically. The distance interaction is explained by the rest mass of photons, which equal to zero.

We know that every human has own photon emission, that called light body, and that interacts electromagnetically with the environment and any other subject in it even on long distance. Photons exhibit dual properties: that of a particle and that of a wave. When coherent, they form a beam.

Photons have a property to refract by the lens and act as waves by themselves. Photons can also appear as a particle when their position is detected.

The quanta, the light particle, the photon, have an independent existence. That is unanimously accepted by the scientists and for that discovery in his scattered studies, Arthur H. Compton won a Nobel Prize in 1927.

To me, it is a light that animates everything in the Universe, in a Measurable Reality. And that light can think and imagine and hence has the power of creation the most mighty power of The One. That light is an Imagination.

Most of the animals don't make their choices. They react to the thoughts of their owners or strangers and usually follow instincts and reflexes that they were trained for.

Insects, however, just obey what the Infinite Conscience implies for them, like the bee that flies to get a pollen from the flowers and brings the honey to the honeycombs. The waves that bee receives from the Pure Conscience, are such effective that work with the mathematical correctness.

70 (68) "The God inculcates to the bee : make you a home at mountains, at the trees and where they (people) build 71 (69) then feed yourself with the fruit and go my ways". From it's (the bee) gut there is a drink comes out of the different colors, that is a remedy for people. True, there is the sign for those people, who think! –The Qoran Sura 16, (Russian Translation by Krachkowsky).

So, it worth to trust to the ideas, that comes from the God, or from the Imagination when we are tuned to the certain state that implies making right choices for us.

Our emotions usually indicate if we are on the right tunes, as so does the intuition.

Sometimes it's thanks to, but sometimes sadly because of, we develop our emotions, out thoughts waves, frequencies and zones of waves throughout our childhood and they are formed under the influence of people who surround us.

At any stage it is possible to switch our antennas to the right tunes and everyone is able to do that.

Just by wish of every good thing to happen to ourselves, to those who around us, to all who is far away and everyone we come across with.

Even when we are in a low mood.

Actually, low mood is perhaps a mechanism to heal our mind, or one of the healing mechanisms. When we are in low mood, we on a low frequencies. We analyze past events, we process the thoughts that are leading to the low mood. Correct attitudes will help those blues to turn to creative tunes rather then to the depression, to avoid mistakes from happening in the future. But what is certainly true is that we all need a rest time to time and getting in low mood is the kind of the rest for us, too.

Remember to stay tuned to the good wishes and intentions throughout your healing process.

Do not analyze too much – don't get into the influence of someone else because the ways of your thoughts that lead you to the low mood or sadness, perhaps was triggered by another person. Wish the very best of your feeling to happen to that person – protect yourself from their influence.

These wishes shall be embedded into our thinking pattern at the instinctive level in order to protect ourselves from us in the first instance.

9.

Thoughts

12. Beware of many thoughts,
because some of them are the sin …
49. Rooms, The Qoran

"Watch your thoughts for they become words.
Watch your words for they become actions.
Watch your actions for they become habits.
Watch your habits for they become your character.
And watch your character for it becomes your destiny.
                    Lao Tzu, Chinese Philosopher, 5th century BC

Thoughts become intentions. Therefore, if someone experiences suicidal thoughts that usually cross the mind of people with certain conditions, (like depression), which is serious, they intensively become intentions. And that is dangerous.

If we look to depression as to a zone of destruction, or a whirlpool, we can imagine how difficult is to get out of there. If someone has thoughts to harm others, and those thoughts are replayed in their brain that is dangerous, too. In medical terms that called psychosis.

At the same time if we accept that those thoughts with aggressive contents (psychotic content) are result of emissions by others, then we shall accept that the schizophrenia patients are exceptionally talented individuals who are able to read thoughts, intentions of others and see things, which others are not simply able to see, not able to recognize. But we all receive such signals.

Thoughts are the sound waves and they are energy. It's my belief that everyone in our world falls under hypnosis of his or her thoughts for the different time, for a second, or for years. We generate our thoughts in our brains, or we might receive the though waves from the external sources and may be aware or unaware of it.

When we are aware of someone's thoughts, if we have such a sensitive receptors that allow us to read someone's thoughts then we might have schizophrenia. Every normal (average) individual received thoughts signals, as they are sound waves of a negligible dimension, not able to be aware of, to recognize those signals by the rational brain.

If we do, then our mental health worth have checked.

But whatever the thought, which becomes a wish, we process in our brain, we fall under the influence of it – under the hypnosis. And only we decide for what time spell we remain under hypnosis of the idea, the thought. If the frequencies of our thought waves correlate with the idea that hypnotized us for the moment, the power generated by such resonance can be mighty.

Same phenomena occur if the thoughts of millions correlate. While throwing ideas into the mass media, governments or groups of individuals often use these phenomena in many ways. Repeatedly throwing relevant information in cycles, mass hypnosis easily achieved by such groups.

Can we measure the thoughts? Modern medicine can assess the activity of the brain by EEG machines and also MRI gives us a picture of the some brain activities.

EEG is an electro encephalogram, a recording device that gives us a picture of coherent or incoherent electric waves.

Neurophysiologists – the specialists, who measure and investigate electric brain and spinal cord activities, group electric waves the brain and neurons produce to alpha, delta and omega waves, respective to the waves amplitude, frequency and length.

Electric activity of the brain, the spinal cord and of the each neuron (the nerve cell) triggered by our thoughts. Neurotransmitters released upon our emotions, which are also triggered by our thoughts and number of other stimulus, like smell, taste, visual perception etc. Neurotransmitters are peptides, which affect chemistry of our blood and organs. Some of them are toxic, some of them therapeutic, the others are neutral to the cells and organs of our body. Thoughts conduct our emotions in their majority, and the instincts, the reflexes and sensory perception also stipulated by our memory.

The purpose why the neurophysiologists look to the brain activity is to assess the nervous system after an injury, to follow up patients with the epilepsy, to look if there any tumor in the brain, or to follow the activity of peripheral nerves, for example, during the surgery.

Sometimes, assessment of the electric activity of the brain is used for the research purposes.

MRI – magnetic resonance imaging is a device which gives us a picture of brain activity, being mainly used to identify the tumors n the brain, the structural abnormalities and to assess blood vessels of the brain and so on.

I'm not aware if there are any device that can measure our thought waves exist in any laboratory, but as those, I believe, are the supra-sonic waves, they are measurable and they obey all physics laws having all properties of the sound waves to comply with.

So if we produce the thought waves, they emitted, travel, they are received, transmitted, processed and reflected back to us. Like an ultrasound wave does. Like every sound wave does.

Such low frequency waves create vibrations.

Accumulated vibrations compose corridor-like structures in space and time, or circles, or tunnels. We fall into them, hypnotized, thanks to our thoughts we replay in our mind, our attention, and our electromagnetism. When we are tuned by our thoughts to the certain vibrations, our mind (and brain) gets polarized to the accumulated tunnels of the relative vibrations and hence gets attached to them: to the tunnels of vibration of joy or sorrow. Every bit of information that attracts our attention can create the tunnels/corridors/whirlpools/

meridians/circles/zones – you name them, in physical and non-physical space. Once trapped in such tunnel, if fell under their hypnosis, you'd find difficult to get rid of them.

I hardly understand if the thought's sound waves possess electromagnetism per se, but it's true to say the thoughts attracted to the certain meridians/tunnels/circles of vibrations when we particularly replay them. Those accumulations of the vibrations I call meridians/tunnels/zones already have certain events chains those encoded into them.

For example, the circles of affluence and wellbeing have already occurrence of awards, contracts, certain meetings, right people in consequence that will bring the individual who fell within those meridians to desired achievements. Meridians of despair, disbelief, bad emotions and insult contain respectively undesired outcomes and events in a row.

Let think on how the activity of our thoughts can affect our lives and lives of others, even though we had never pronounced our thought. Yes, our thoughts are active when we replay them in our minds.

Every live cell has intelligence. Every quark has intelligence. That intelligence derives from the immeasurable reality, that creates everything and even before the creation has all details. Thus, it's (intelligence) processes the information, including the thought waves, received by the quark, by the molecule, by the cell and so on.

When I say before creation that really doesn't apply to the immeasurable reality because the time dimension does not exist for the source of everything, including the time dimension.

It has created the time dimension to keep an order for us, but it doesn't comply with any of the dimensions it has created, including the time.

Hence signals received by the quarks and the whole objects, instantly processed, reflected, transmitted in a physical reality and answered to that signals in an

Immeasurable reality, or in other words, reflected by the Immeasurable Reality, modern quantum physics calls it Supreme Consciousness.

When the individual emits waves of the idea, thoughts, wishes, that waves meet the waves of other people and zone of interference are generated. In those zones at the levels of mirror reality, measurable reality and immeasurable realities, new ideas meet, reflected, resonance and bring occurrence of the new solutions and events into their existence.

They create sparkles, they correlate, and they interfere. It would be possible to observe such tiny waves interference once they are measured and their existence proved experimentally.

Same phenomena occur when the individual creates such waves and sends them around (daydreams). The waves reflected by the physical objects, including our own body organs and the DNA, mirror and immeasurable realities and the reflection area generates zone of interference, too.

Thoughts and wishes can be created in our own mind but they are also travel in three realities.

Within the zone of interference new ideas and sparkles are born, depending on zone, destructive or constructive ideas fluctuate.

When we all become aware of the impact the thoughts may make to our health, in my view no disease would remain uncured. But if caught by any disease, let professionals to help.

So, before refusing or rejecting any treatment, one shall think: with the Iwishyou attitude your waves are more likely to cohere with the waves of the idea of treatment, the team of scientists and pharmacists stands behind the treatment, the doctor who looks after you and your family who wants you get better. Remember, physical waves, once generated, recorded in mirror reality and immeasurable reality and exist there forever. So when your waves cohere with those of which brought the drug/the treatment for the purpose, achieving your healing is more likely, because the intent would be coherent with the zones of interference, the Iwishyou attitude waves will work and your cells will pick up the intentions of healing that already exist in the powerful constructive interference zones.

Within the zones of interference constructive ideas and ways of approaches to the problem, solutions will immerge spontaneously. They already exist, but actually come at the right time with reflections from physical objects, from the level of mirror, and from immeasurable reality, where solutions already exist.

When the problem occurred, it occurs together with the solution. Brain can recognize the solution in the certain states. The best training to turn the brain for the ability to recognize solution is either to meditate or to tune onto iwishyou (positive) attitude.

Iwishyou is a state of mind where there is no place for the anger, for rage, for the envy and resentment. They are usually taking our attention away from healing, from constructive ideas, ability to generate effective approaches and to create happiness.

But realizing that we reflect, receive reflection of our own thoughts and wishes and also receive the signals from others makes me wonder, where this knowledge would take us?

Again, everyone would choose their own direction and here is a very important factor that the direction's layouts would have been installed by the person's circumstances from their childhood. The repetitive scenes we witnessed during our childhood often have programmed our emotional status. Whatever they are, we might find difficult to overcome that emotional status as we feel comfortable in it, even thought if it is far from our natural state- the state of happiness.

This will not apply for those who grown up in a happy families indeed. The child, who grown up in a happy environment, been placed into the tunnels of positive energy already and naturally, would succeed in life, being ready to overcome the stresses successfully.

But everyone will gain significant benefits by applying the knowledge, the theory of which we've developed hitherto. To do so, it is important to remember: our natural emotional status is happiness, by wishing every good thing to happen to us and to everyone around, we send the energy signals to ourselves and creating electromagnetic fields of prosperity for ourselves. There are zones of interference to remember about, too.

Isn't that beautiful? By our sincere desires to wish every best thing to others we do not only create a better interference zones, but also by reflection, we receive those best wishes to ourselves.

Remember, I'm not suggesting to get into a direct contact with every stranger on the street, I'm talking about the management of your thought's navigation, about the control of emotions, about developing the attitude and hence the base for the clear fields of happiness and developing the great healing power we all naturally posses.

I always fascinated by how the Persians, Tajiks and Uzbeks call the God. It's actually a Persian name, "Huda".

The literal meaning of 'Hud" is the Self.

So those who say we do create our own reality actually are right. Because our wishes actually, being the physical carriers of the energy, the sound waves, literally create things, the events and bring people into our range. The Universe is arranged in a way, which makes our wishes we wish to others, to receive ourselves.

We release the wishes and thoughts from our brain and first, in the brain itself cells reflect and react to the thought waves by immediate release of neuro - transmitters.

Second, every other organs tissue cells receive, analyze and reflect those waves at all levels from an atoms of the peptides to the whole organ level. The consequences of the neuro- peptides released and their action to the cells is brilliantly described by an animation in a wonderful movie, The Secret, part 2. It's available on Amazon and if you watch it, it depicts exactly how the cells of the body react to the chemicals released into our bloodstream by hippocampus in response to our emotions. But first, I believe, what happens is that the cells, tissues and organs receive the wave signal of thoughts and reflect them, process and store in the DNA (or activate related parts of the DNA during the processing). Changes of the chemicals in the cells, organs and whole body just follow, catalyzed by the neuro- transmitters.

Third, while we release that thoughts wave outside our body, these waves to be reflected by the Universe, by the physical subjects of it and by the Supreme Intelligence.

We emit possibilities and opportunities, whether we generate them ourselves in our mind or do get them from the outside. However, for the God, in whose Imagination we all exist, there is no "outside" notion.

Sometimes I get fascinated how individuals make up their world they accept as the only reality and confine it into their stipulations, created by themselves. Those stipulations are mere their thoughts and imagination confined by the lack of experience and knowledge. While we are, the doctors, acknowledge that as a natural process, as the individuals feel comfortable in the environment of beliefs they have been introduced to, it is important to accept that the way the Universe is functioning and the properties of the Quantum Field are immense, infinite and impossible to comprehend. For our own best, we do not need to know everything and understand exact mechanisms of creation in details, but the basics.

It will allow us to stick to the way of achievements, the way of wishing and receiving, appreciations and happiness.

If there is a problem, there will be always the solution. Trust it exists and you'll get it in right time. Wish every best to everyone around; pick up those thoughts and frequencies that make you happy and joyful.

10.

I wish you

5 But for those who gave (to those in need) and feared (to harm) and reckoned the truth is the most beautiful – will get the easiest to the easy (from The God) ...

92 The Night, The Qoran

Wherever is the problem, there is a solution.

Always. That how the nature laws operate. Wish get better for others, your wishes will reflect to your mind with the solutions options just activated. The solutions were there for you always, but your mind was not able to see them. Wish the better chances to someone, ask for your best wishes to come true, and even better will manifest for you, a sparkle of the great idea enlighten your mind in the moment, your "iwishyou" attitude becomes your natural state of your mind.

Wish, connect, communicate, throw around your best intention and get them back from the organic cells, non- organic subjects, from the mirror reality, from the immeasurable reality, where uncountable opportunities already exist. Wish to your family, to your friends, to those you love, to those you hate, to Yourself with the same intention and same faith and sincere.

Whatever the zones of interference you are in, I'd suggest you to keep general safety rules, like keep an eye to your child and not allow to get into a contact with the strangers etc.

Watch your thoughts, your words, and your intentions. Art, books with the beautiful words, the music, the nice scents are the best inspirations for the delightful thoughts and hence, for the therapy. Many therapists also suggest keeping a pet that also works brilliantly to take your attention away from the stress.

Those patients, who irresponsive to the conventional treatments may benefit from the training their brain and tune on the attitudes where treatment is possible. But no one shall abandon his or her treatment. Create and let yourself into the constructive interference zones.

Some people are naturally optimistic, generate lots of positive emotions for others but even they, when get into the destructive interference zones, may get dysfunctions like depression and other serious diseases.

We all face stresses in our life. But in order to evolve we have to come out with the circumstances we face. To protect against falling into the frequencies of despair and sadness iwishyou attitude, I believe, can work as a perfect tool and avoid the destruction and the influence of the energy generated by the negative thoughts.

The zones of the negative and zones of positive thoughts energy also have an attraction power, thanks to their electromagnetic properties, but constructive zones attractiveness is much more powerful. That derives from the intrinsic to the all creatures desire for happiness.

The direction where we all evolve is happiness. So it's important for us to know what is the happiness for us, and wish that to our family, to our friends, neighbors and the strangers we come across. I'm not talking about talking with every stranger on the street, neither about donating money you can't afford (however, donation itself is a very powerful tool, too) but I'm talking about building the ability of the mind to constantly build, generate, emit, recognize and transmit constructive thoughts, ideas and waves that would be a ground for the prosperous individual and the society.

Some would say it's impossible to trace every thought, but it's worth to remember how thought may transform our body, causing damage to it and sending despondent signals to whom we love or to strangers around us. Let your sorrow thoughts to get away, wish them very best.

Your wishes of best, the beautifulness you throw around shall work for you in the first instance. They also can bring strength, make others aware of the potential and create zones, where ideas of fulfillment the potential emerge as the sparkles.

Where damage has already occurred, that manifest as the dysfunction of multiple organs and systems for the long time or depression, the tool of training a "wishyou" attitude will help to overcome the disease but might take some time.

In some cases, I think, the positive change may happen overnight (but would require the maintenance and support), but in majority of cases, where damage at the cellular level persisted for many years, the healing with the developing of the attitude can take months. That refers to the healing of the damage occurred either to the physical or the mental health.

To get the treatment, the first thing to do is to see a doctor. There are hundreds of diseases people suffer from worldwide. However, epidemiology of the diseases vary in different parts of the world.

Also, there are as many medicines to treat the disease or to control the condition.

If you are religious, just think that God that sent you a disease has a cure for it, and not believing in that is a disbelief in God's power.

Some of my patients say they don't believe in doctors, they don't believe to pharmaceutical companies, they don't trust to the NHS but all those thoughts are again, the waves, we create ourselves that bound back to us and our bodies. As a result, we suffer form them ourselves and make those around us to suffer.

The worst part of our disbelieves is that they are become true as we tune on them; we catch those thoughts we are tuned on and thus create a whirlpools of the waves that affect our cells.

Let me at this point to tell what I sincerely desire for you, if you are a sort of the patient who doesn't believe to the modern medicine. Wish you healing, controlling of your disease, wish you to trust your doctor, to trust the pharmaceutical company, the whole team of the scientists, researches and developers of the medicines who stood behind each drug, yet to earn their money, they'd wished to treat/ control your disease, trust the NHS, they really do fantastic job and every staff personnel is a human who'd wished you to get well in your life!

Always remember: every disease requires attention of the doctor; that is a rule of thumb. If it's not serious, the doctor will tell you.

For example, there are many people suffering from the chronic conditions, like high blood pressure, diabetes, asthma. Some patients with such conditions controlled by modern medicines doing quite well. But others have their conditions poorly controlled by the same medication. Doctors all would say the reason is why it happens poorly understood. My version of the reason is the individuals poorly responding to the treatment tuned to the frequencies, I call despondent of fell into zones of destructive frequencies. That is not someone's fault. We don't have sufficient knowledge of such zones and waves, our thoughts create but everyone would agree that the main causes of the low moods and diseases that add to low mood and state of depression are not related to the personality: sanguine or cold- blooded, we all can fall into depression. Our immune system suffers from depression, and range of other systems would do, so.

There is past experience of that treatments from previous generations and truly, most of the medical treatments work very well. Trust to your doctor, get the treatment and continue to train your brain to tune for the constructive waves interference zones.

There will come an era, where no disease would exist as the humanity will learn to use it's incredible natural power to heal and restore it's health to the pristine condition but as we are yet to accept, embrace and learn more about it, in my view, using available options to treat the diseases is the most appropriate way. Where the damage had occur on the cellular, tissue or organ level, certain mechanisms needs help to start their (cells) new life. I think, nowadays society is not yet ready to understand and fully use the potentials of the curing mechanisms and conventional medicine should work. In some instances that would work as a placebo.

There are certain mechanisms of the disease, which well known and medicines work to overt such mechanisms. But if the individual were turned to the tunes of distrust or lack of will to cure, even powerful medicine wouldn't be that effective as shall be.

For the individual who's full of willpower to achieve a cure, even placebo would work miraculously. Again, that is not to blame in any mean those who irresponsive to various treatments, no. If the person trapped in between the "bad effect interference zone" and just a sensitive to the thoughts that person can't feel, but the organism receives, processes the information, than we can have a clue to the wider range of knowledge of how disease can overcome the body.

Then that mean each of us is responsible for the disease of the individual we don't even know. We all are The One, aren't we?

Destructive ideas and thoughts are the sound waves, which accumulate in zones and, unfortunately, like a whirlpool, can draw in. Patients with depression are example of those who drown inside the whirlpool of bad waves and destruction zones. They are not able to get out of the situation by themselves and often require a professional help, indeed.

But constructive ideas and zones of the positive waves have even greater electromagnetism power because they are more natural to our initial state and main propose of our existence.

In my view, it is logical to assume that, provided that our thoughts are the sound waves and they vibrate the media, according to the law of the energy conservation, where no energy disappears, our good or bad thoughts are accumulated in zones of identical vibrations. Then the whole world is the vibrations. Everything consists of vibrations and affected by the influence of vibrations. At the end, the every quark is the sound and the vibration!

Perhaps, if we clearly know about existence of constructive and destructive, negative and positive, good and bad vibrations accumulative zones, we would keep trying our brains stay tuned to the good, to the constructive waves.

That is for the sound wave. It travels through the air over time, but does it takes time to it to travel through the immeasurable reality? I think it doesn't take any time for the wave to travel through the reality, which I call immeasurable. That reality doesn't comply with the time-space dimension and I return to the explanation of my understanding of it later.

Every thought is a wave. Every intention, every wish is a wave. We emit the waves, receive them, reflect them, attenuate, process, transmit, send and receive the reflections. When intentions, wishes and thoughts are scattered, we receive their scattered reflection. When they coherent, we receive them back in a coherent fashion. When our intentions, desires, wishes and thoughts don't violate those of others, we receive them harmoniously focused and they correlate with the waves emitted by others: by the people, by the non- organic objects, the organic tissues and cells, including our own cells, and by every atom and it's components.

I was graduated in Uzbekistan, in 1997. By the time I've started my graduation journey, which is formally 6- years long curriculum, I was living in Soviet Union, but when I finished it, not only the whole country has changed, but the Science over the whole world was re- shaped…

Different approaches to the understanding of etiology (the causes of the disease) and pathogenesis (the mechanism of the disease) has started, new powerful drugs developed, new diseases and problems have emerged.

Now I'm living in UK, with my family, work in London and my close friends from childhood often repeat telling me that I was always wanting to live in London because seen it as a capitol of the Science.

Science in the air in London: it's everywhere here. Every child here, as in whole United Kingdom, knows lots of aspects of the science and the Science is behind everything here. Everyone gets lots of knowledge here whatever subject they want to learn very quickly, thanks to the Science, Researches Institutions and ordinary people who work hard to develop them in London, in Manchester, in Birmingham, in Cardiff-everywhere in UK.

Ordinary people in UK highly appreciate the Science, paying tremendous respect to it. Government, Private sectors and charities plug huge funds in researches and education. Indeed, the Universities of the UK having their place among the tops of the whole world deserve that. Researches usually are very scrupulous, transparent and the research stays behind every aspect of socio-economic, political, information, technology and any other institution.

I found that is a fascinating culture of the Science in the Western world. I also fascinated by the Christianity as a religion, which is considered in the Quran as "the brother of the Islam" religion.

I love the ancient idea, which is clearly derives from the Christianity, that by giving, we receive. The idea to love sincerely, to embrace your enemies, to desire the very best wishes to your neighbor, and to fall in divine love with all people. Same ideas carry Buddhism, Islam, Judaism and all other religions, but in Christianity, the emphasis on wishing the best to everyone we encounter, "Love thy neighbor" idea developed and the mostly unconditionally postulated. And there is a reason for such hypothesize. By giving our best wishes, we protect ourselves from negative states of the mind, from poisonous thoughts and zones of destructive interference. That is a powerful protective mechanism that every prosperous individuals/families/society uses in order to prosper. Every individual shall use it in order to become truly successful, in my view.

There is also a very powerful formulation in Islam for those who reckon themselves our enemies: wishing them all peace.

It is hence obvious that the major proposal of every religion was and still is bringing a harmony in people, their personalities and in the society. Every religion teaches its followers to the virtues that help to achieve harmonious interaction between people, the groups of the society and the followers of another faiths. All religions teach to live in peace, in happiness.

But as some researches showing, the happiness doesn't relate to our faith neither that does to our economies.

For example, Christianity is a dominant religion in Europe. It's also a main religion in American continents.

The research conducted to measure overall Happiness Index (HI) in countries like China, India, Brazil, in European states revealed that Brazil is the country where HI was the highest. Next to it was India. HI in western rich countries varied. I personally attribute that to the culture that dominate in those countries and of course, to the climate features.

According to the New Economics Foundation (NEF), the UK's leading think tank promoting social, economic and environmental justice and The Happy Planet Index (HPI), the leading global measure of sustainable well- being, the Happiness Index (HI) is highest in Mexico, Argentina, Brazil, China, Kirgizstan, Laos, India, Indonesia and Thailand. It is significantly lower in UK, in most of the European states, in Russia, Canada, Zimbabwe, and Australia. HI is lowest in America, Chad, South Africa, Congo, Afghanistan, Mongolia. (**Happy Planet Index: 2012 Report http://www.neweconomics.org/publications/ entry/happy-planet-index-2012-report)**

Happiness Index is measured on a base of how many long happy lives it produces per unit or environmental output.

In natural weather conditions, where we are deprived from the sun in most times of the year, we naturally, experience the lack in vitamin D. That contributes in low mood and blues. In Saudi Arabia usually people spend most of the time in the buildings, and wear clothes which not allow sun exposure to the skin due to extreme hot weather and strict religious regulations. The lack of the sun exposure may be same for the South Africa, but in Nigeria, Chad, Uganda and other African countries, which constantly experience shortages in water resources, torn by war or by poverty and diseases Happiness Index cannot be adequate.

It comes surprisingly unexpected, that in rich country, like United States of America, HI is among the lowest ones (http://www. happyplanetindex.org/data/). European States also having their Happiness Index among the low ones, too.

In European states I attribute HI to the weather, to the stress, overwhelm of information and the culture. We are being entertained by movies depicting the murder and sorts of crimes that don't make us feel happy. The thoughts incited by modern movies get us to the vibrations of scare, horror, aggression. I'd rather prefer the Bollywood movies about love and opera songs for the entertainment.

We all emit thoughts of negativity, and sometimes we do so inadvertently. When thought waves meet each other, they create zones of whirlpools, where, depending on moods, destructivity or elation dominate. Vulnerable individuals pick up bad tunes easily and are sensitive to the effects of zones of interference. Therefore, the cultural constitution of our minds plays huge role in our thoughts vibrations, and hence their polarization. The constitution of our thoughts indeed includes our entertainment's content.

We can guess what others might think, but we cannot know that for sure. We cannot read thoughts of others, if we do, that's mean we suffer a schizophrenia – the disorder, which is poorly understood yet. But I believe, we react to the thought waves and moods, the others spread to us.

By analyzing, criticizing or praising someone's actions, we become a part of that someone's vibrations channels.

However, there are plenty of evidences that we can, by training of our brains, achieve an independence from the moods and toxic thoughts of others and progress to the levels, we program for ourselves.

I 've noticed that everyone who criticize someone's actions, be those in past or in present, who judges, particularly with the resentment, inadvertently falls under the influence of that person.

By paying our attention to the actions, persons or events, we are judging, we magnify our attention with emotions; we empower the vibrations of negativity or positivity. As an example, I invite you to think about the tragic phenomena that occurred in civilized Europe, having its start in 1930 and continued to the one and a half decades of horror. Whole world get involved into the World War 2 at either part of the battle and the horrible ideology of Nazis played it's role in killing millions of innocent people at industrial scales.

For the future generations we have to remember the victims of the WW2, those who passed away, we pay our tributes to those who fought against the fascism and pray for those who suffered tremendous atrocities. But we shall not analyze the ideas that lead the fascists into their existence in order to avoid the fall into those frequencies. According to my theory, the ideas that, once accumulated, create the tunnels of the identic in their nature frequencies and while we replay the thoughts and ideas, we make our minds polarized and attracted to those tunnels. Hence the humankind is quite aged species, such tunnels already exist and are ageless. When we fall in such tunnels by our minds, staying tuned to those tunnels, our ability to get out from there depends on number of factors. Such factors are: the attitudes of our personality, our family background and the environment that shapes our psycho. Whether we accept that or not, by replaying the thoughts and ideas, we fall in the realm of vibrations of those thoughts and ideas. That how, in my view, hypnosis occurs. It is not always the case when we can distinguish whether we fall under the influence of hypnosis or not. We may and actually we are under hypnosis of the ideas, political parties, views of the others, are fans of the celebrities without even acknowledge of our hypnotized states.

When I was practicing in Uzbekistan, I used to see 50- 60 patients a day for the 10-11 years. In my country we work Saturdays, so weekends consist from the Sundays only. There are few public holidays we don't work at: the New Year day, two- three religious celebrations and the National Constitution day. Multiply 300 days of the year to 50-60 patients a day and you'll get 15-18 thousands patients a year. Nowadays in the Private setting in London I'm seeing less but can devote bit more time to details.

Overall, I can tell you that was many times when I faced the patients in different stages of their disease and those diseases varied from the nuisance ailment to the serious illness. I can clearly understand from the face expression, from the intonation, the body language whether the patient suffers the chronic anxiety or whether a normally relaxed individual experiences mental shock because of the scaring diagnosis.

I would not describe here stages of the shock or personality disorders but I can clearly tell that the persons I have ever came into the contact with exist in their reality created by their mind and their reality, their ideas of those realities, their beliefs and wishes are the major component of the outcomes in treatment they get.

That is not a try to place the burden of responsibility of the treatment outcome to the patient in no way. That is my understanding of one of the aspects of the success in treatment. Fail in treatment is not the patient's fail; it is our fail, of the medical team, the doctor's and that of the patient. If the patient would have know that their beliefs are the major mechanism of the healing process and if the patient knows about their abilities to navigate to the vast range of realities, they will get better outcomes. It is ours, of the medical team that includes psychologists, responsibility to educate the patient that there are choices of the realities we may choose for ourselves, those have already been existing in a range of possibilities which we are to choose ourselves and that we are not rigidly attached to those we have established in our perception of the reality. That our perception of the reality we currently have is usually confined by the past experience and the stipulations/ ideas we have received from the external objects, and often hypnotized in that confinement. But our reality can expand to the range of our imagination. If we are able to imagine something, then there is a possibility that we can experience that in our life. The only condition is to get into the tunnel or zone of the joy and happiness- our natural state. In our natural state we are more likely to bring that event in our life, because the situation is already exist in those zones of affluence and freedom state, all you need is just to replay the scenario in your mind. The opportunity will turn over- you will recognize it.

To program yourself into the right tunnel entrance, promote best wishes to everyone you encounter, and for your family, to yourself. Your brain will then start to recognize details you were unable to see and within the right vibrations will get used to make correct choices.

The situations around you will amend, your healing processes will speed up, your brain will do it's job: not only within your body but in outside environment, too.

That is true to every of my patients- they fall even in deeper despair if someone from their family or friends repeatedly points to their mistakes, to the wrong perception. And it's true for the psychiatric patients – they never accept there is something wrong with them! Healthy individuals accept they might be wrong, but even for the healthy person continuous experience of same damaging effects would bring to the wrong perception, interpretation and misunderstanding of the information around.

In some cases our environment or either our perception of it actually might work as a start of obsessive-compulsive disorder, bipolar, depression, dementia.

What happens to our environment? Is that thing going wrong around us or is that our mind that develops rigidity in our perception of the events and people or do we loose our Self?

Do we always develop rigidity ourselves or do we fall into the influence of others or do we simply fall in zones of hostile interference of vibrations? Can we always control our thoughts and protect our mind from the hostile vibrations? Are our believes that build on our experience are always relevant? Can we change their direction of our thoughts and attitudes? How are our believes, that turn us to rigidity, formed? There is a matter of repeated attacks on our brain by ourselves over time, by our negative thoughts, our fears and anxiety.

Let me explain how I'd advice my patients to deal with the stress of some common but very difficult problems.

My remembrance of the past hurt me. As a child, I was neglected and abused.

I would suggest you to speak with that child, you used to be, as an adult, who knows that child best of all. Simply close your eyes and imagine yourself as a child. Tell to that child, what would you do to him or her to make that child to not get upset, your sun will raise and you are beloved. Beloved by You, Yourself...By the God. By those, who surrounds you today.

Best time to do such exercise in my view, is at early morning, while you are still at your bed.

Because echoes from the past disturb us mainly during night sleep, and if wake in that mood that we caught during night sleep, that mood would often catch us for the whole day. Those are parts of the normal life cycles, we do not need to scare of them, but we can deal with if those are causing a problem for us.

The Person in front of me abuses me. Take a few seconds and think: I'm likely to fall in anger, but there I make a choice whether to take that as an insult and hence to abuse the person back,, or to protect myself from that person's bad vibrations. Do not analyze the reasons, do not seek any explanation of what is happening: abuse in any form is wrong. Do not get involved in those bad vibrations the abuser currently in; that is not an area where you want to be! We are here to be happy, at the end. I wish them back all the great things to happen, inside me and tell that loudly outside. That what I usually do when I meet someone wicked.

You have experienced a heavy loss. Every one in his or her life encounters a loss of someone very close. That can be our parents, other family members. But guilt, prolonged bereave, depression are not what those who had passed away would want to see you in. Remember: no one gets away vanished. We all are going to meet up again. We can talk each to other. As per the law of energy conservation, we are, our thoughts and our souls exist forever. That is the energy exchange, we are transformed from one form to the other, but our love to each others eternal. Don't get too deep in bereave, don't make upset those who passed away. They feel you. Wish them a piece, the very best you might wish, talk to them in good mood. I sometimes make cheerful comments and joke to my Dad, who passed away when I was 17. I can't hear him back, but I want him feel happy.

My wishes that born in my mind, pass my brain, pass my blood, my cells and the DNA. They would vibrate the certain parts of my cells and the DNA, the certain parts of my brain and make it work ahead or disturb from working constructively. It occurs to everyone while he or she encounters wishes born in his or her mind. Choice is yours to replay and process every scenario.

If similar situation is happen to repeat at any time later, and I'm controlling my wishes, desires and thoughts, what choice I'd make? If the situation repeats over and over again, how I'm likely to react?

Conclusion? Wish beautiful things to yourself every morning, to your family, to everyone you meet throughout the day, to strangers, to colleagues to every single person in the world, to every flower, to every animal, every soul – they all have a live intelligence to reflect your wishes back to your life!

If feel any illness – get quick to your doctor! Modern Science has learnt most of the diseases, there are plenty treatment options available, get them!

Feel emotionally unwell? Visit your therapist, talk to the relationship specialists – they've been taught to resolve the problems in relationships, go see them and wish them every success in what they do! Even if you feel low, even you don't trust to the modern medicine, let try, you might be surprised by the outcome.

I am always skeptical to any new treatments, theories and methods.

As a teenager, I was very skeptical towards religious postulates. But what I certainly convinced now is that the religious books are, indeed, much more scientific then I used to think about them by the age of 19.

My hard studies in the Medicine and my past experience made me convinced into that even more.

I'm not a religious person at all, but I read religious books of Indiums, Christianity, Islam and Jewish Torah, as a child. Indeed, I would have many questions about certain statements, the answers to which would come decades later. I know now all books of all religions are Science encyclopedias encrypted in messages intended for the wide public and lay people.

The knowledge all religious contain been gathered for ages. It is sad that some of political camps use religion to achieve their control and power using dirty strategies, misinterpreting the religion and using the lies.

I wish everyone to understand those dirty politics have nothing in common with the religion. I wish everyone in the world to navigate their wishes towards own prosperity towards the world where no place for the lies would remain, there will be no necessity for that.

There is lots of words and phrases devoted to the thoughts and clear definitions of required virtues in the Quran, the content of the Wisdom of Solomon is that fascinating with clear description of nature laws in the book, that each time I revise them I wonder what is religious extremism have with the religion in common at all?!

In Hinduism, there is a statement that Buddha is everywhere: in every creature that animates the flesh, and in every single atom, which is nowadays in the universal law of the physics. Modern science is full of proofs that those theories are true. Christianity delivers the most important message to the humankind: love thy your neighbor-the greatest protection and tool to evolve ever, but everyone finds in every religion what he or she seeks. Everyone navigate to the vibrations where he or she wants to be.

I invite you to think about the purpose of life. To me, the purpose of live is an evolution. The evolution towards better in way of happiness. If we ourselves decide what is happiness for us, we'd probably achieve it at some time. But can everyone exactly know what is the happiness for him / her? Depends on that person's potential. How long it takes to us to discover our potential? It can take whole life or just a few moments of childhood.

For the reality, that created us in it's imagination, that is a matter of no time to know our potential. So it's very important to keep connected to the source of our creation to know our true potential and hence, by filling it, to discover what is the happiness. Actually, that is a quotation from the religious books: You may get annoyed from the things arising at your journey but you don't know why they are happening.

There are some brilliant ways to connect with the source of creation through the meditation or I'd suggest to read books like of Deepak Chopra: "Creating an affluence", where the reader invited to pay attention to the 25 qualities of the quantum field, "7 Spiritual Laws of Success" and many more

We all have an ability to evolve which is inherent to our existence. Think about the simple protozoa bacteria. All it has is a single cell as a whole body / organism. It doesn't have a head, nor does it has a brain. It lives in colonies (in a clusters, like a community) and whether we accept the fact or not, it evolves.

We are, doctors, face the fact on a daily basis that bacteria becoming resistant even to very powerful antibiotics. Where does the simple microorganism get such an intelligence or wisdom to evolve? It is intrinsic to its existence.

I think, that the fact of existence of the small creature is equal to the fact that it has a power to evolve in order to survive. Some insects develop mutations during their lifecycle in order to adjust to the environment, as well as some animals and all that because they evolve.

I personally believe that same occurs to the humankind.

I encrust my DNA with my dreams and wishes to see a better word, to live in wealthy beautiful environment where there will be a great place to every single person.

Watch your words, because they are your words. You use them, and they will reflect to you. Watch your thoughts, you might think are yours. But while they might be yours or not, you choose which of them to process, because you are tuned to them. So we are constantly process the thoughts. Where they come from? Whether they are coming out from ourselves (from our brain cords), or from others or directly from the Supreme Reality, we choose to process them or to leave them, or to stay tuned to such sort of thoughts.

We cannot read or hear thoughts of others, we can feel the general radio-magnetism of the though waves, that what we call our intuition. We can only guess the content of someone else's thoughts, but we can't know them for sure.

If we can hear or read thoughts before processing, then we are having condition, called schizophrenia.

Everyone with this condition requires doctor's consultation. Because only doctor will be able to assess if someone who hear or read thoughts (and that is quite possible, they really do), only doctor can decide if the person carries any danger to others. In old times, such people would stigmatized as witches and be badly punished for their paranormal abilities.

They really hear and read thoughts and what is the fascinating in Psychiatry, (what my psychiatrist friends tell) their patients really able to read thoughts to see unseen to others and even that some other people actually can develop such abilities. Science is yet to explain those phenomena to us.

Not only our thoughts can be processed and reflected to ourselves, but also they would reflect to our children's DNA, as well as our wishes, our intentions and our deeds. I think that might relate to the DNA of our other relatives, too.

Thoughts are the sound waves and comply with the physics of the sound waves. Sound propagates through the media and vibrates its molecules. Provided that ever our cell contains the molecules of our DNA-the most important part of every cell in our body, propagating thoughts vibrate the molecules and the quarks of our DNA.

How the contact between the same DNA carriers formed, the Science is yet to explore. But it is true that if something good or bad happens to our relatives, we receive the information through our dreams and feelings.

When Science proves that the God has created us by his imagination, and we all exist in it, what a great new era would start!

For now we may remind ourselves that we have the imagination gift, too. That is a great power that needs to be used with care.

11.

Anatoliy Kashpirovskiy and mass healing phenomena

It's true and quite possible!

It was the end of 80ties when the new mass healer emerged in USSR. Everybody would know him, Mr Anatoliy Kashpirowskiy, a qualified psychiatrist doctor.

Every single person in Soviets heard about him and irrespective of their trust to his treatment methods, Mr Kashpirowskiy was known to everybody in Soviet Union. Even skeptics, who had rejected the phenomena, when he was first introduced on TV in the USSR, will watch him whenever he appeared on TV live. There were thousands of people gathering at stadiums or around the scenes, where from he was speaking on the stage, purportedly hypnotizing the people in front of the stage and those who watched him on TV.

My Dad was alive then, my Mum is still remember those days, but she doesn't remember what was happening to her during those shows sessions, when we were watching on TV how Mr A. Kashpirovskiy was hypnotizing people in far Russia, ordering them to get their diseases and ailments healed in variously wicked fashion.

We were watching the program in Uzbekistan regularly live on TV.

Both my parents, Dad, a vascular Surgeon and Mum, a Gynaecologist, were very skeptical about the purpose of mass healing. Actually, we didn't believe that is possible. Academicians argued about the ethics, surgeons ridiculed the idea of effectiveness, comics earned enormous popularity in mocking him. But each time Mr Kashpirovskiy started his translation, Mum, like hundreds of people among thousands viewers sitting in live theatre and watching Kashpirovskiy, would fall in a sleep, sitting and twisting her head around her neck (that was normally always aching) with the speed like her head was a helicopter propeller. When he stopped his speech (usually long philosophy monologues), she'd awake. She'd able to repeat, if we've asked, every word she herd, and she would remember everything, but she would deny anything like twisting her head at all.

Each time I would wonder if her neck is aching at all? Only when I reminded her, she'd start to complain on her neck ache, however, that very year she'd forgot about her neck and spine problems forever. Different diseases started to trouble her much later.

Not everyone would react in a same way. I was a teen, about 15 years old, I would only feel a tears coming out from my eyes, while I was watching, but I was contributing that to the wonderful music running at the background of the Hypnotist's speech.

In my view, the phenomena of mass healing could be explained by the theory, where thousands of powerful antennas altogether were emitting the waves of intention to get healing. They all fell into the hypnosis of the single idea- to get cured.

If I just imagine the undertone of thoughts of thousands in those stadiums, millions in front of the screens during the real time live translations and created by such emission frequencies (all coherent!)

that makes me think that everything could happened to the people who were there in that amazing atmosphere and concentration of power, emotions, positive thoughts, intentions and their electromagnetism. For those who entered those tunnels of wish to heal, healing of everything would be possible.

Before they have attended the sessions they have already firmed their stable intentions to cure, indeed. All were having a hope; otherwise, it is difficult to explain huge audience gathered at the shows with the tickets price being massive for that time.

Every person who'd get into the stage for the healing their personal ailments by the Hypnotist, would receive wishes to cure from all those who were watching them and all looked quite impressive.

The magic happened almost every time with almost every patient on the stage. Critics accused Mr Kashpirowskiy in "makeover" acting.

The audience was desperate for cure and indeed, if follow the theory that thoughts are sound waves, and intentions are accelerated waves, in that situation, everything was possible. Thousands of generated emissions created powerful electromagnetic tunnels of hope, trust, love, adoration, glory, might ... Remember: intention and deep desire add power (presumably, electromagnetism) to the emissions of thought waves.

Mass hypnosis is possible using the ideas, which are misinterpreted. Same dynamics work for mass madness. Mass madness, for example, was the international politics in Europe in late 1930, when the European civilized countries obeyed a group of mad people allowing the Nazis to kill half of the world over years, kill innocent people in industrial scales for no reason!!!

So, mass hypnosis can be used in different ways but if the individuals keep hypnotized in their own ideas of affluence, health, good wishes to surrounding people, love thy neighbor attitude, then I think the whole world would change to the prosper environment.

I doubt if everyone would trust to Mr Kashpirovsky even on the time of witnessing the phenomena of mass healing that would occur in real time. But even they did not trust, they would accept the fact of extraordinary experience.

The theory of the wave nature of thoughts, focused and enhanced by the immense desire, would explain the phenomena in some extent.

As the thought's dimensions so tiny, it is reasonable to assume that they can affect the approximately same sized particles, the most fundamental bricks of our bodies, the quarks. I think the interference may occur in no time, as the quarks are so tiny that require "negligible" attention, as the modern physics states.

If you were convinced in that the thoughts are the sound waves and might affect your body, would you allow yourself to think and replay every thought that comes into your mind? Knowing that thought can interfere with your particles in no time?

It might sound awful to the patients with the Generalized Anxiety Disorder and add even more to their condition, but the theory states that the thoughts themselves are of a negligible dimensions and if not replayed regularly, their effect can extinct by the effect of other thoughts.

At the end we have to accept that we are who replays our thoughts in our minds and we are able to manage them.

I personally don't believe in that the only certain people have power to heal others. Everyone can become a doctor, provided that the person has academic abilities and able to study in university.

I also believe in successful treatment using clinical hypnosis. Everyone can learn the technique and use it. The most fascinating aspect of this fact is that every sensible person makes own choice whether to fall into any sort of hypnosis or not. It is my understanding that at the end, everything in the material world is hypnosis (of the sound) and everything is a result of hypnosis.

I believe that while an evolution is intrinsic to the life of any form, the humankind, the person, an individual can choose the direction of own evolution. And every individual, in fact, does.

We choose the direction, where we want to evolve to. Wrong thoughts, some sorts of the waves, space of bad frequencies can lead us to evolve to the monster killers, but others can bring us to the significant wealth or to a Nobel Prize award or to the records of the Guinness Records Book.

We can evolve to the horrible despots or to the beautiful poem – writers. We can also to evolve to both, but here is the catch. While one can deceive others, the individual cannot deceive own brain.

There will be individuals who'd continue to propagate bad frequencies in order to achieve personal benefits: wealth, power, and so on. But those shall think if they want their own bodies to receive signals of damage, disease, and their children to receive reflection of waves, they've sent around.

I personally know people, who achieved incredible wealth, lived extraordinary lives and died, leaving gravely ill children behind. That make me think: is that their DNA received the signals?

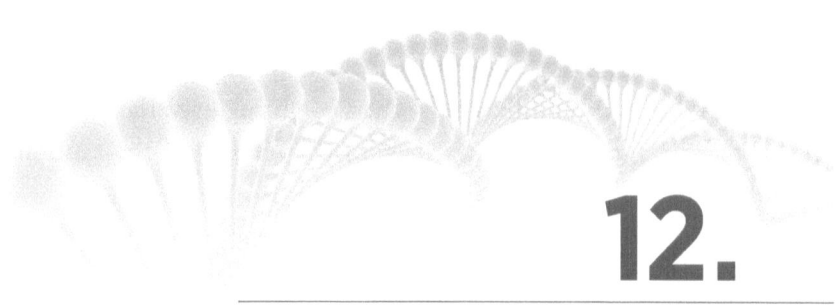

12.

The tool

"There is no path to happiness: happiness is the path."
- <u>Gautama Buddha</u>

An ultimate purpose of my book is TO INCREASE AWARENESS OF HAPPINESS. We all shall live in happiness; that is our natural purpose.

During our life journey we come across diseases, raises and falls, and those are natural circles of our evolution as the seasons change during the year. It is important to keep our natural state we are born to experience: the taste of life, celebration of life, happiness. We are here to bring up next generations in happiness, to love previous generations, to give care, love, passion, and sympathy.

I use diagnostic tools to find out what is wrong with the person such an ultrasound scan, which is completely safe for use. It is equipment, where the transducer emits the sound waves beyond the hearing threshold, and all organs with different frequencies reflect those waves and the machine transfers that into the image on a computer screen.

The liquid, normally contained in a bladder, or bile, filling a gallbladder, which is also a liquid, reflect the sound waves with the different to the soft/ hard tissues intensity.

Every organ, its tissue and cells receive, transmit, reflect, attenuate and propagate sound waves. They do so with the different extent, but certainly, every organ does so. I believe that there is a direct link between the cellular DNA of all organs and the brain and the speed of that interaction is very fast.

In our DNA, we have thousands of genes, where variations of the disease/ health /conditions perplex and twisted all around and between each other's, like a jigsaw of the combinations of the letters. And those combinations are switching and linking among each others accordingly. How the functioning of each gene occurs is not yet thoroughly understood, as some of our genes start their functions before our birth, while others at later stages of our lives. There are also thousands of genes, which may keep dormant and not activated for the whole life but may be inherited by our children and grandchildren.

It is true to state that the high functions of the brain are the most mysterious for the modern science. But there is established association between the stress and higher morbidity. When we upset, in sorrow or anger and rage, our brain processes the combinations that create the words/thoughts of disease. Even thought scientifically we call thoughts information and nowadays we give a different scientific formulation of how the brain processes the thoughts, we all accept the link between negative thinking and the various diseases.

I invite you to think about thoughts as the sound waves that carries tiny amount of energy but they still able to vibrate similarly tiny particles of our cells.

It is, indeed true that processing the information, contained on our thoughts by our rational brain also contributes greatly to our health and immunity states. But the words and the information as the sound waves that mechanically may affect our molecules and hence functioning of the organs and the tissues/ cells is the theory which I want to study more about. Otherwise, that would not possible to explain in full the link between the stress, continuous bad mood, released bad chemicals in our bloodstream and diseases like cancer, because no such chemicals could be found in blood when the disease have already established. We can only find the markers of the small tumors, if we look for them at the beginning of the disease.

However, high blood pressure is the condition, in which systemic release of stress related catecholamine (with lifestyle, genetic profile) may play its major contribution.

Whether that is an interaction in a waves fashion or by the neurotransmitters from the brain to the DNA, the science is yet to discover, but my message is to tune to the thoughts of perfect health, to dreams about things you would achieve, and those whom you love would achieve and note every beautiful detail around you: a flower, a pet, the scent of the spring, light touch of your baby, a smile. Never think of any harm to anyone in the entire Universe, those thoughts would harm you in the first instance.

I wish you – is a tool. Like a physical exercise for the flesh body, this tool will enhance your ability to tune to the vibrations of "good" frequencies. It will help to create those frequencies for other people, too. Every day, every second …

I believe that it will help people to overcome diseases, to help treat depression, speed up other healing processes. It's nevertheless not to substitute any other kinds of treatments, but to accomplish conventional ones.

Let's face the fact: we are not only physical bodies. Let's accept that. What some refer to our spirit, or light body, that animates our physical entity, I would refer to the Conscious, which includes the following features, but consist not only from them: our Intuition, Our Emotions, Feelings, Awareness.

My understanding of the life is that there is a field, science calls nowadays a quantum field, which is a source of everything, creates whatever it wants. I would refer the quantum field to an immeasurable reality, which is infinite and hence the real, more even real then our physical/material, measurable, constrained by space- time dimensions, reality. We can touch, feel, see and measure it, move within it. When we are entering immeasurable reality, in order to exist there, we would loose our shapes, image, appearance and the whole physical body. Because that reality is Our Imagination and it is The One.

I think that every quark coming into existence in our physical reality comes out from the immeasurable infinity (that must be a source of the form of light) as a light particle in wave fashion in a position of probability. It constantly moves in the material (measurable) world placing the proposition of it's probable amplitude as a trace in several locations. And manifests in a certain location when it meets attention or the place, where the most frequencies correlate, coherent at the same space/time event.

That is literally means the quarks are everywhere. Occurrence of the events are everywhere, meeting the right people are everywhere, our healthy life or the disease are everywhere, all we shall do is just to choose the time and location by placing our attention

There. And there is a possibility that already exist.

For instance, if there is a group of people believing in an occurrence of some kind of event, that will emerge from the countless probabilities of existence, potential of which is already exist in space/time environment, they are likely to meet that event. Same power would have a single individual. If you imagine a picture you'd meet in your life, you will see it one day. There are countless possibilities ate the same time point, you choose one or fall into that which is chosen by someone else. So, the group of people have only to create an opportunity by sending the signals (the thoughts waves) and by doing so, they make the space where generated currents amplifies, interfere and bring into existent the quarks of probabilities into the certain location. The only stipulation is the sort of vibrations where intentions of the group of people or the individual were played in. There are "good vibrations" tunnels, where all good events occur, and "bad vibrations" tunnels, where only bad things come into existence. We choose those tunnels by our thoughts and intentions.

Why we have religious conflicts in our world, if all the religious teachings point to the same things?

I believe that happens because political interests use religions as a tool. People are naïve enough to swallow 'religious' causes for the conflicts but there are small groups of individuals who earn bloody billions on such conflicts. Such groups of people create images in the brains of millions who by believing in what they see, helping them (inadvertently) to create new wars, gain more political power, rule the world.

Wish to every quark, to every atom, to every person and every galaxy to manifest in position where they can be happy and emit the thoughts and desires of happiness, complacency for everyone else around. Then, your thoughts waves will be received, processed by the immeasurable reality and reflected by it, also will be reflected by very quark, by every cell of the person in front of you, by the Universe from the limitless skies, back to you by turning every current into an opportunity for your wishes that will work for your best.

Does the Supreme Imagination attenuate or accelerate our thoughts? While transcending from non- material reality to the material world, and wise versa, the waves have to change their polarity and hence their properties. Physical becomes non- physical; the fact turns to the probability, the position and location to the possibility. That is very interesting question, I wish I would have know the precise answer. I assume that the waves, attenuated by physical objects, are enhanced in reality, which is beyond time- space characteristics of the physical world.

I'm fond of the idea that everything emits own waves of a many types (electromagnetism, light, sound waves) and so say the Science. All those waves correspond to the energy.

If my understanding of the wave – nature of existence and interaction is correct, then the particles are emerging from the Imagination (non-material) reality, which is immeasurable, that dreams and in it's imagination creates the world we call material, which constrained in time- space dimensions then reflected and submerge again back to and from the source.

The mirror- like interface between the physical reality (which is, to me, the product of imagination of the God, where we all exist) and non-physical reality runs at every single level of life existence: the quarks, the molecules, the cells, macro-organisms and galaxies.

Remember to pay attention and hence think about those things you want to happen in your reality, wishing everything better to everyone around you.

Nowadays many psychiatrists and therapists advise to their patients to look after a pet as part of the treatment for various conditions.

Do not wonder: it works. Not only looking after your pet takes away your attention from the stress, it gives us vibrations of love, adoration and wishes of well-being.

Animals are the best example of receiving, reflexing and behaving in respond to the thoughts and intentions, but the problems occur if the poor creature was repeatedly attacked by bad intentions or even actions of it's surrounds for the reasonable time. Its brain would unable to refuse bad frequencies and it would behave in a direction of harm, misery, tragedy. We know many examples when dogs unexpectedly become aggressive.

I believe that the research if conducted would give us a better knowledge on how the dogs may react to our thought waves.

I want to invite you to think about the interesting phenomena I witnessed with my student mates in 90ties. In Samarkand, where I was studying in Medical Institute, we were doing our rotations cycles at different hospitals. Each rotation was between 2 to 4 months length. When the time was to have our rotation in oncology hospital, I was amazed to see dogs and chickens near hospital, which lived at neighboring houses in surrounding areas, and were having tumors bulging out of their bodies ...

I 've asked myself if the tumors are infectious and how they could get to the dogs, living at houses near to that hospital and even to the chicken? Professor oncologist, whom I also asked the question during a lecture, Dr ...., has replied that there are many theories about the cancerous tumours genesis but eventually, no one still know the exact mechanisms of etiology (the cause) of the disease.

He pointed, however, that the majority of our colleagues- oncologists who used to have had sorts of the tumours due to the radiology treatments, they were giving to the patients, but that was an era, when relatively safe radiology departments were not available.

Next to the oncology hospital was located Maternity hospital No. 3. Around it there were no such an animals suffering from the tumours.

Now, many years later, after the shock I 've experienced by seeing that phenomena (pity, I did not had a camera by then and the cell phones were not available in early 1990s to capture them) I'm still baffled and still thinking about the mechanisms of spreading of the disease to the animals living near the Oncology hospital in Samarkand in 1990th ...

Are there certain vibrations were accumulated, to which the animals were prone to get into?

Now, when I came to the assumption that the each thought is the wave, I think that every organ, receiving those waves, absorbs, transmits and reflects them. Even the liquid processes and ultrasound waves, so why the organs can't? Why shall we think that the tissues of our brain and other organs do not receive and process the waves- the thoughts?

But here we are to think about health, prosperity, and happiness.
I want to invite you to think about some facts and to find solutions.

Attending to regular conferences in London gives me an access to the discussions among medical doctors: why treatments of some diseases do not work equally on all patients? Why would some respond to the treatment very well and why the others not only feel improvements but also develop wide range of side effects, even previously unknown ones, from the treatment?

Sure, not every medicine would that effective, as the pharmaceutical companies would want them to be. But even in cases when well known treatments provided for the diseases we know pretty more about nowadays in all medical specialties you would hear that the attitude of the patient, their emotional, spiritual state, the mental health are the major contributor of the outcomes we have as a result of the treatment.

Does this wonder you: why the treatments in wealthy country, in UK, where the best medicines in the world are available, where the hospitals are equipped with the novel sophisticated equipment would not work for everyone?

Perhaps, that is a modern life style adversity, but low mood, under recognized depression, poor sleep, sedimentary life style, pressure of too much information- all contribute to morbidity and it's outcome.

Our attitudes, our unhappiness and sometimes despair are the major causes of the situation with our health we are currently in.

One of the prominent biophysicists of our time, Dr Marco Bishoff, compares us to the antennas. I'm fully agreed with him in that. His research has shown that we can catch the ideas, the thoughts, which travel in space like the radio waves. According to him, we choose the matter to which the content of the thoughts is related. In other words, we would tuned to, and hence, receive and process the thoughts and ideas, we want to.

Once I met him at the conference in London, I asked if he believes in potential of self-cure from the cancer. Without hesitating, he replied: I do. However, in my view, people need time and efforts to comprehend that. More research need to be done.

The Science needs more evidences and relevant research may prove it is possible.

Like antennas, we receive, transmit, reflect, remit and create waves (thoughts, wishes, daydreams), which are highly likely to be a form of the sound waves. Our wishes, when emitted, reach others, reflected by the Consciousness, the bodies of those who surround us and those whom we want our wishes to reach and don't want them to reach and reflected by every single atom, the molecule and every single cell. When come to our mind, thoughts and followed moods, I think they are not merely born in our minds. Some of them visit us from other sources. We are to understand them in future.

Good thoughts and good dreams are much more powerful than the bad ones, because they are constructive and are the sources of creation.

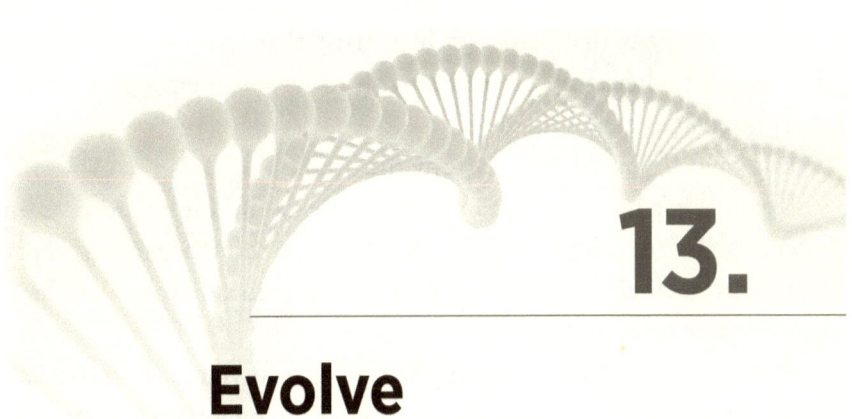

13.

Evolve

"If you truly loved yourself, you
could never hurt another."
                    - Gautama Buddha

I genuinely belief that the imagination is a great power that evolved a simple bacteria to sophisticated macro- organisms, like a humankind and animals.

I do not suggest that the bacteria have had such an intellect; the simplest forms of the life are rather manifestations of the Intellect, exhibition of its creativity. But even though the simplest form of the living micro- organisms did not have an intellect, it has had evolutionary properties in order to survive, to overcome adversities it has had a power of the precise new designs deriving from The Mind. I 'll return to The Mind and it's manifestation later but let think about the daydreaming. Dreaming in order to evolve is very peculiar to the very nature of every organism, every cell.

It is a power of our dream that has brought us into the place where we currently are. Our emotional states while we create our dreams are very important. On our very emotional state depend the outcome of scenarios we paint in our mind. On wishes and desires we encounter while create the dreams, I think, we define the time of manifestation of them. Whether that state is a firm trust that what we dream will happen or whether that is a sincere wish to have something better for others intensifies the outcome.

But what if you do not have any dreams?

If you don't have any dream, you might be in low mood or even in depression. You might feel you are under the influence of negativity from others, which is usually very easy to get rid off from. You have already learnt how to do so from this book.

But what if you do not have any dream because you have attained all of your dreams? Then you must be enjoying the happiness of existence or you feel there is a final stage of your journey?

Then, in my view, you would have to generate the new ideas, new dreams, because that is a natural force that moves us forward and is very peculiar to every single creature – the gift from the God we all own.

Dream for yourself, dream for the others, use your gift for someone for your own sake, because that powerful mechanism we have, if unused, if stagnated, may destroy our lives and that of our beloved ones. You will learn how that can happen once you experiment to develop iwishyou attitude.

Evolution is our endless journey, the purpose of our existence. We have evolved from the single cell into the sophisticated machines that has a power of thinking, dreaming and hence creating. Why would we evolve? What is the purpose of the cells division inside the ovarian tube while the blastocyst moves towards the uterus and settles in the deep lining of the womb? The single fertilized egg moves towards the womb in order to grow, to evolve, to become a human being, to enjoy the natural state.

The natural state of our minds is happiness,

In that state we can move forward, we can dream, we evolve - each of us, all of us.

While some individuals, families, groups of people, societies, or even nations fall in despair, poverty, which often lead to a tragedy, others are affluent, prospering. Interestingly, the HI (Happiness Index) of the wealthy countries is not higher than that of the developing countries.

The emotional background, the vibrations and specter of the thoughts of individuals affect the ability to respond to the stress.

Happiness doesn't refer to the amount of money each individual or nation posses but is affected with the environment, the Science has yet to name.

The state of the happiness, however, might affect our wealth, physical and mental states and health. I would refer that environment to the vibrations that exist in the field of the Consciousness, in which we all exist and that is more then the space, time dimensions. There is a field that hosts whole life and Universe that cannot be measured and Science has referred that as a Quantum field. It's indefinite, infinite and immeasurable; it's the most primordial form of life, where noting is exist but yet it is a source of Creation – Pure Consciousness.

Where our disbeliefs come from? From the disappointment?

I noted an element of a huge hypocrisy in people representing religion who don't actually belief themselves in the innocent conceive of the Jesus Christ. It is a blasphemy to tell about such hypocrisy but I want to make my point clear: I personally convinced in fact that Jesus Christ was born from the virgin.

In the Quran, its written: (31) when Maria's mother (Wife of Imran) asked the God to give her a child, then, after giving a birth to a child, she brought the child to the temple. "Here is she, my daughter." But only God knew whether that is a daughter or the son. Because the male gender is not same to female gender. Surah 3 "The Imran's family" Lines 31(35)-(36)

Why? Have you ever herd about individuals called intersexual?

There are lots of the articles about such anomaly emerging nowadays. It is condition, where same person would have organs of male and female genders. There were cases in UK and in China, when adult male patients with blood in urine would be investigated and discovered to have the uterus and ovaries.

There is nothing that would convince me that St Maria couldn't have properties of both genders. At the time of her puberty she could be able to develop both gender cells and conceiving of a baby is quite feasible in those circumstances.

I am not a religious person, I do not pray every day, but I use the power of wishes to get everyone every best as the tool to throw around good vibrations. I always fascinated how all religions teach as to the same, but everyone gets what he or she seeks even in religious books.

It is my belief there are also the cells in our brain that act like the cords of the musical instruments, they give a birth to the new thoughts, new ideas.

We decide how to react to them subconsciously and we decide on how to precede those thought, born in our mind or received from the outside, consciously. How that happens, depend on the individual circumstances. Whatever our decision is, thoughts, while replayed in our brain, in other words, when they are processed, are making changes to our cells. Be aware of that. I would recommend to everyone to remember that the thoughts have the power to change our bodies and environment. Dreams, when we day dreaming, created by ourselves, emitted to the air/space and being spread to the world and reflected by it.

Therefore, it is very important we truly wish all best thing in the world to even those who we think our enemies, because being existing in the same environment, we will inevitably receive parts of that waves to our lives, too.

Our positive, I would say, constructive thoughts, whishes, dreams and intentions are much more powerful then destructive ones. Once released by someone, they do not only protect us from destructivity but amplifies our growth and prosperity by creating relevant vibrations. If you grow a flower and sing to it every day, you will se how flourishes it. But without songs it will rise and flourish in a less extent.

Intuition – that perceives the waves and gives us a feeling of what is right and wrong for us.

I strongly associate the Intuition to the logic and past experience but we have some feelings even without previous experience.

Emotions are reflection of the thought waves. Our Intuition is the sort of feeling and emotion, too.

It's depends on state of the mind : turbulent, or relaxed, how the waves would be processed and what emotions they would entail.

The areas were thoughts are positive, which are, in my mind, realm of vibrations, give us an impulse, an idea, an inspiration and take us up, elated. But negative areas attract to us events, circumstances and people we don't want.

Those are areas positive or negative thinking, I'm not talking about positive or negative emotions, I'm talking about the thoughts. They lead us to relevant vibrations area, like polarity of the electrons and protons.

If the theory of thoughts nature as the sound waves is true, whether you accept it or not, their interference do exist and you cannot get out of the natural laws of a life in the Universe.

Modern researches have shown, that an abusive behavior triggers release of the bad catecholamines in the bloodstream. However, the same release occurs when the person experiences bad emotion being the subject of abuse (upset, sorrow, anger) Those catecholamines together with the other poisonous agents make their damage to the brain.

It is unfair that not only the despots end up with having serious brain diseases, bad sadly same may develop to their victims, too. Dementia and many more mental diseases, which, in some extend, develop due to periodic release of the poisonous neurotransmitters to our brain.

When we abuse others, our body creates chemicals, which destructive to our mind. It occurs even when we abuse ourselves by thinking of worthless, falling into the negativity about our self-esteem and ourselves.

But when we allow others to abuse our feeling, then we let their negativity penetrate our mind by processing the information we received by abusive phrases, there we hurt ourselves. If at that point we return abusive words and wishes by wishing very happiness to the offender, we defend our antennae from the damage, our soul from the whirlpool of negativity. Our wishes, remember, reflected by our DNA back to us and at the same time protect us. They reflected by the DNA of the offender opposite, too, no matter where he or she is. Whether that is a cyber bullies at the other part of the world or our office colleague or a horrible boss, for the Imagination – the Creator, it is of no matter.

Every bad word, and even inadvertent wishes, will reflected by the DNA of the wisher, of the speller of offensive words to their DNA, the cells of the hypothalamus WILL produce destructive chemicals; damage to the cells of the brain of the offender WILL occur, bad waves of the intention WILL emit to the brain tissue, to the vibration aura, to the Immeasurable Reality and to their destiny. Alas, that happens every time. Unless we acknowledge that and change our thinking, we are unlikely to achieve lasting youth, perfect health, luxury conditions. It is also important to not fall into the destructive frequencies of our abusers, to not undermine our self in our own mind, to not take abusive words to our attention. In other words, don't trust to any word of the person that offends you. Take as the regret to that person, who doesn't know what the damage he or she makes to his or her own brain, health, and destiny.

You don't have to allow those bad waves to penetrate you mind, your Self. THINK!

Think for your own good, wish the very beautiful things to happen to your offender, think of every good thing, and do not allow the negativity to your vibrations. You are in your own vibrations that create your unique reality, and you are only who builds that reality for yourself. Should you wish to live in vibrations of happiness, wealth, perfect health and appearance, lasting youth, emit those vibrations, create your ideas and stay in them. Live the full life as you want that to be for you, allow those vibrations and words you want to, do not pay attention to those words and vibrations, which you do not want to disturb your ideas and tunnels you are in.

Do not assume that if the person (who just offended you) opposite is wealthy, she or he is happy, because truly happy person will never offend anyone around.

Can we harness our anger? If we do, we are protected and not fall into the frequencies of bad aura/tunnel/vibration channel, whatever we call them.

Can we defend from the poisonous thoughts and desires?

If we truly wish the happiness to our enemy, we are actually doing that. Their bad intentions reflect to them, and not harming you, if you do not allow them to your mind, if you don't process them in any way. Reflected by your defense vibrations/tunnels, those bad words, wishes and vibrations hurt the speller/wisher/creator of bad tunnels, but who knows, may be by suffering they'd grow, evolve and get to the true happiness by change?

Or let's imagine you are wishing your enemy the greatest happiness to occur to their life, but that enemy's greatest wish is to "dance on your grave". Just wish him or her happiness, they'll face a grave if their desire was so intensive and deep, and only God knows who's grave that would be.

Same effect carries being at frequencies, where we attract bad thoughts and emotions or reflections of bad thoughts and ideas, emitted by ourselves or by others.

Stress is the natural event we come across throughout our journeys in order to evolve. Child doesn't grow without illnesses- that how the immune system adapts to the environment. Each baby comes to the world, crying. The way we respond to the stress shapes us and hence our associates, particularly family members as their DNA is the closest to our DNA matrix.

What I'm seriously believe is that, while evolving individually, we moving towards the world, where no non- curable disease would exist, where our beloved ones will be available to talk with and see and share life even after death, where organs will be available without any need to sacrifice someone's part and there will be no any conflicts and no need for money to be mentioned at all.

For some that might sound ridiculous but for the scientists that is achievable and mechanisms of such evolution of the society, while understood, are yet to be named and used.

While scientists are getting to their achievements in their laboratories, I'm inviting you to achieve higher HI in your own life. I suggest you to exchange the most beautiful words and best ever wishes to people you come across with. Then, provided your thoughts and wishes are sound waves, receive those from others, even from those whom you don't have to know.

Pay attention to everything that triggers your better feelings: that can be a classic lyrics, your own words, parts of the books you enjoy or quotations you heard from someone. Wish to a stranger or to your beloved one, or to whom you perceive as an enemy/competitor the very best wishes you want your beloved ones receive and for yourself.

By giving, you'll receive. Each day you receive good wishes to happen in your life, you need to silently, or verbally, but sincerely, wish same good things to everyone you meet on your daily routine, be those your family members, colleagues, mere passerby, a pet, a tree, a flower, every live creature and organism. You will see the results within the one months of trial.

Conventional medicine implies taking the medicines, conduct investigations and in some cases, arrange a surgery. Here, it's worth to mention that there thousands of staff recruited often to create, to develop, then to test the medicines. Researches conducted, trials tested and medicine reaches the chemistry. Their role is to fight the diseases. So, conventional medicine is not to be refused, but we also need to tune onto the healing waves ourselves, too.

Those naïve people who'd think that their thoughts and intentions is just an innocent moment and would not reflect into their lives or that of their families or that of their beloved ones always wondered me.

Remember: whatever you think, whatever you wish, is the physical waves and they will be reflected into your life by every single atom, by every single molecule, cell, micro – and macro-organisms inside you and all around you as well as be reflected by the whole Universe, irrespective of if you want it or not, whether you accept that or not.

When I was a medical student, we had to spend few weeks' tutorial cycles in Psychiatry, and attended the psychiatric hospital based on Isaeva Street in Samarkand. We were medical students from different backgrounds, but what we faced within the hospital was the state of the patient, which is difficult to describe and the behavior of some of the medical personnel, which were difficult to understand. Most doctors and nurses were forgetful and nervous, ALL were in state of stress, reflecting the stress, living the stress, emitting the stress. I wondered if that is remnants of the Soviet era, where low funding for the hospitals were in blame?

During the soviet era, when any religious idea and hence belief in hidden things were not allowed, some of the staff used to abuse their patients. But others told us stories, when patients could read their exactly thoughts, see the events from the future and many more that would make me think that hallucinating patients simply see things that normal brain is unable to recognize.

Actually, by the time some of the medical and non- medical staff would develop abilities to see unseen, too.

Those who abused the patients, would receive thoughts and perhaps, hallucinations of the aggressive content, but those who loved their job and were empathic and treated with passion, sympathy, mercifulness, would develop thoughts and ideas of better content.

Then, the question would be raised for the medical student: who is the patient and who is in fact, receiving the treatment?

I know a very experienced psychiatrist in London, who told me that there is phenomena known to the scientists, indicating that over the time, people that come into the contact with the patients with schizophrenia who hallucinate may slowly develop almost similar hallucinations, but doctors usually don't tell that to others for the obvious, ethical reason.

That makes me think that those people who looked after the patients simply developed their ability to see subtle things that was previously hide from them. BECAUSE WERE UNAWARE OF THEM before. But subtle objects, scenes, which in normal world we call hallucinations existed even we were not seeing them. Some people may develop the ability to see hidden things with time. This ability to see hidden things is uncommon in normal people (let call us average brain).

I remember my Dad who was dying from undiagnosed by then condition.

Albeit he was looking fit, was very strong, a very tall man, (2.20 cm height), he suffered diabetes mellitus. He suffered peptic ulcer, but since his diabetic neuropathy made him to feel no pain, he developed internal bleeding and died from the complication.

We, my Dad, my aunt and me were in one of the rooms (in the library), of our house. Dad lied on the sofa, feeling bit unwell. Suddenly he deteriorated. We called the ambulance.

My aunt was older then my Dad for about 16 years. She used to tell us many times how all the siblings and parents adored my beloved Daddy, the very youngest son in their family. She many times recalled what is the mischief child my Daddy was in times. So, the time of his childhood remembered for the whole family in details.

When my Dad has suddenly deteriorated, he started to hallucinate. He saw many people in the room, he described them to us, asking, how can't we see them?

For me, a teenager, who saw my beloved Dad as a strong man, a doctor, seeing him distinctly hallucinating was such weird, that he was not joking, was seriously telling us that there are lots of people, the soldiers in the room and many others, that I couldn't belief in what I was seeing, I was scared. I couldn't know he is falling to coma.

He described those people, called some by names, told to my aunt what they were wearing …

Suddenly my aunt burst in tears. The ambulance arrived by then, they did not allow me to go with my beloved Dad, I stayed with my aunt at home.

She couldn't stop crying: "He is dying"-she whipped. I shouted on her: my Dad is a strong man, why do you think he'd dye?

She replied: "Your Dad named and described a relative who died 10 years before his birth. That relative (she named him, but I don't remember the name) was buried in cloths in which he died. And your Dad explicitly described the same cloths on him".

She was crying in such grief, like as My adored Dad has already died and he did in the very next hour, but, of course, I was not accepting what she is saying at that moment ...

What do you think? Are the hallucinations odd or real? There is no doubt, they are real for the person who sees them. If they are real, then, can you imagine what the person, who constantly sees them, lives with? Can you imagine what that person lives if those hallucinations are of aggressive content? I f they repeatedly surround that person, and make him or her upset? And all that happens against their wills, because they simply see or hear others, either who had passed away or what live people play in their minds?

It is true that the person who is hallucinating and is tuned to the tunnels of negative, aggressive, bad vibrations may impose danger to the public, but for many of them it is impossible to make a comprehended choice.

The other true is that nobody can tell for sure, if that person doesn't sees/hears content of the other's minds or able to see/hear things that exist below the sensitivity of the receptors of ordinary people.

In my phantasy, I'd think that everyone would accept the patients from psychiatric departments as special people. In some patients who don't impose a danger to the public, their abilities could be used perhaps in research or in other areas. Although they are very susceptible and often vulnerable, there are many facts we can learn from such patients, in frame of the legal ethics. The realm of psychiatry is steel full of surprises.

I wish for all of the world population to become aware of the power of our thinking, and even tiny amount of us create positive vibrations tunnels, no place for the negative ones would remain. No other people, neither psychiatry patients would be involved in negative tunnels/areas of thoughts sound waves.

It is true to say that the "happiness in the air" atmosphere is directly linked to the attitude of the people, of the community, of the country. Everyone can feel that. I regard that feeling to the vibrations, people emit by their thoughts.

Compare the 'air, the atmosphere' feeling of Brazil, Indonesia, Russia, India, US, Italy, China, South Korea and UK and in each town or city there. You feel the atmosphere of emission of people's thoughts, their mood, which is often not related to the economy of those countries. It depends rather on the media that spreads the speculations and hence rules the moods, or sort of movies people watch (in Western countries). Turning to the Eastern countries, like India, Japan, South Korea, China, the atmosphere of the vibrations depends on the spiritual beliefs of people there. And also, the culture in most Eastern prosperous societies doesn't support the entertainment by watching movies that depict murder in abhorrent details.

Even country, city, neighborhoods, have their own frequencies and hence feel of affluence or misery.

Same applies to the different families.

14.

The Sun of the Family

We are the result of our thoughts.
"You are what you think. All that you are arises from your thoughts. With your thoughts you make your world."

Gautama Buddha

Every mother is a Sun of the Universe called family. All other planets: the Earth, the Moon, the Jupiter, The Venues, etc. (the father, the grannies and children), turning around it. Ideas, feelings, thoughts of the mother would embed in children's lives in significant extent. The mood of a whole family, kids in particular all turn around Sun's emotions and thoughts. The course of each planet lined by the Sun, as a life of the children, because the ideas, feelings, thoughts of the mother would embedded in children's lives in significant extent not only during the course of the pregnancy but even in later life. They would reflect by every single cell.

In one family, you feel that "happiness in the air" atmosphere but in others you feel an impending tragedy, sorrow, depression. And that is often not related to the wealth of the family.

What is the major contributor for the atmosphere? I believe that are thoughts in the minds of every family member. Child's mind is very sensitive to emotions and I believe to the thoughts of their parents.

The children naturally often are not aware of their abilities to see more that the adults. Because they don't have experiences to fulfill their prejudices and their conscience is free to see and feel more then we do.

In happy families and in unhappy families there are cycles of good and bad moods. But there is a tendency of bad things to occur in unhappy families more often.

In families, where constructive frequencies and vibrations are filling the space, destructive thoughts and tendencies just disappear.

Beware, however, that every family member can fall in bad frequencies from the outside world, particularly those who are vulnerable.

I wish that every family member aware of the theory I'm talking about and use it daily as a protection tool.

In unhappy families thoughts come and repeatedly visit a person, but because the frequencies created in unhappy environment are destructive, the likelihood of the bad things to occur is slightly higher than in happy families.

Even if one family member feels bad, every other will feel that, too. In my opinion, the DNA of the relatives sends signals (thoughts) released from the same DNA owners that reach each other's even faster and effectively, then any other form of communication through thousands of kilometers! Otherwise, how to explain the fact that if something happens to our relatives, we can receive that information through our dreams in same days even without even the phone ring or Internet connection?

To me, everyone receive such information, but not everyone pays attention to it.

Same DNA holder's thoughts waves become coherent more easily, enhanced together and attenuated together, in my view. That makes the stressful event harder to survive, because add to the frequencies own "negative" emissions. But with the sincere wishes of all best each to other, the families likely to overcome any difficulties with enhanced strength and creating new potentials if the emissions that are "positive".

We cannot blame each other's in creating bad frequencies but shall indeed create our own, to help others and create a better environment for ourselves and for whom we love. That is why we communicate, share, empathize – these all are important aspects of the successful treatments.

In happy families, raising the children, good beautiful words, the sounds, the flowers the fruit- all reflect positive thoughts that are the permanent attributes.

Personality formed under our emotions and thoughts throughout the time and starts to form during our childhood. Our personality determines our destiny and lifestyle. However, we are able to change our personalities, and hence, our lives. There are simple steps to do so.

The first is to accept that we are here for the happiness. Then to define what is the happiness is for us. Third, get into the tunnels of prosperity, affluence and creative, positive vibrations. You already know, how.

You may wish to share a beautiful dream or a song's lyric or music that came into your mind, be that an opera sounds, or any other beautiful music, a poem, a sound.

I love "Strangers in the night" of Frank Sinatra, or Marcus Vianna composition or many, many more. The brain would choose itself the healing music notes or a poem lines or a person whom would like to speak to. The brain would tell you what works as a healing frequencies each time it falls in potentially harmful frequencies. It is important, therefore, to recognize what makes you sad, or feel unwell, what thoughts to avoid, what company to choose.

But always remember to help whenever possible to whomever possible. If someone in a sad situation, don't waste your energy to find the culprit, but help to the person who is in a difficulty. Give love, support, make feel better, feel better yourself.

The main point is that you / your brain would develop the ability to recognize and attenuate the currents you /your brain receives from outside and enhance the potential to bring them to the levels, where only constructive emotions and intentions would live and where the stress would, if faced, never feared but handled in a best way, where there will be no place for despair but for the happiness and the "bliss"

– the buzz word nowadays. It means awareness of our existence – the happiness.

That is the natural state for the Homo sapiens. If we able to maintain our natural states, then the accidents or any undesired events we face with in a daily life would deal with more sufficiently.

I'm sure, there are millions of us, "wishyouists" who sincerely desires everything beautiful, wonderful and amazing to happen to everyone we encounter during the day, every day. But here I'm talking about robust intention to create a better place for ourselves by wanting that to everyone around us EVERY MOMENT, to turn that to the natural state of our mind.

Those who gather in the churches, mosques and the temples wish their best there, they know what is the atmosphere of the wishyouism is. They know how the most beautiful thoughts of bystanders make you feel enlightened, elated.

That would be achieved by making the mind pay attention to the words of beauty, happiness, seek the relevant information, put in writing, record that in the brain and make a WISH. How intense is your wish determines on how quickly you achieve the state of true happiness and your brain matter starts to emit constructive waves / currents.

Brain will learn to the state of wishing best things to everyone around, until that becomes habitual state to the brain.

Individuals don't have to suffer in order to evolve, however, for some of us, that is an evolution: to experience some sort of circumstances, events, to meet the different people.

What we read, the movies we watch, the news in the media – all we perceive would be better distinguished and we develop the ability to choose whether to fall into the bad vibrations or not with our clear awareness.

When the whole world knows that actually the information traps of some groups or individuals who play the terrorism or other wars there will be no need to play such tricks – no one will believe them, neither follow, no money such groups will earn.

In wishyou world wealthy elite will spend all money to gain attention of goodwill – the money will loose the power they own nowadays. The only currency that counted would be intentions and the ideas.

All the ideas we entertain our mind with, are also experience that lead us to our personal evolution.

I don't claim that all your life will change miraculously overnight when you start apply the "thoughts are the wave" theory in your life. I suggest that developing a robust attitude "I wish you all the best things to happen and I'm sincere" to everyone you encounter in your life, to the whole world, to yourself, will increase your abilities to resolve faced stresses, to approach the obstacles constructively, to pick up new positive ideas in your life, to use the best opportunities, which are there.

The opportunities for everyone are right here. We just need to pick them up. We are all in One imagination world, where everything is possible.

If we have those desires, there are objects for our desires and there are simple ways to achieve them. Cautions about the desires: always remember wishing the sufferings to someone will reflect to you, to your organs, to your cells and to the DNA of your relatives.

Unfortunately, there will be vulnerable and mentally ill individuals, but when technical progress will allow us to recognize villains easily, they will be treated.

I know, there will be a time in our world, when there will be no place for diseases, there will be no death, we will be able to invite, to see and talk to our ancestors, everyone would own a planet, and cloning will be taught in schools. Poverty and restrictions will be a history. I know that because the nature of all creatures is to evolve, we are doing so and we will to continue to evolve. I wish everyone to experience that.

I remember how my grandmother used to describe the haven: there will be enough to snap a finger, and whatever you wish will arrive to you straightaway. I was a small child, there were no computers at that time widely available, my grandma didn't know about the existence of the computers and I didn't know.

But that feature of the haven my old Granny herd from her parents often comes to my mind when I click my touchscreen on the android, ordering earrings from India or food from the Tesco.

There were either theoretical knowledge of the imminent evolution or intuitive understanding of it, or both. But my logic tells me now that there even better heaven before us. Because we don't stop to dream – that is our natural power, we shape and re- shape our dreams and hence, we create.

If only quarter if the world's population would become aware of our abilities, accept and embrace them, they will start to wish every beautiful things to happen. The power of their vibrations will change the world rapidly.

Think, some conditions are nowadays potentially curable. Few years ago there were no cure for the vitiligo – autoimmune skin disorder. But now it is treatable. I know many patients, whom I referred to the Dermatology Clinic in London (to Dr S. Chopra and his colleagues), those get their complete treatment from vitiligo. One day, the multiple sclerosis would get cured, too. We have almost won our battle with the cancer – remember the T-cells medicine discovered and developed in UK recently, in 2014. The AIDS also seems to be treated at early stages, so hope is there would be cured and eradicated completely.

That would be the physical influence of the wishyouist people that would change the whole world.

Everyone will be a truly wealthy, cloning will be done by pupils at the schools and everyone will own a planet.

Now is the time to the step up to attain naturally better future, to create better ability of our brains and send correct signals for others, for everyone of us. Create whirlpools of great wishes, beautiful desires, genuine ideas and peaceful thoughts. They don't have to be about money, but about the true happiness coming from the deep inside of our awareness. For some of us, our work brings us happiness while for others, just spending their money. So let wish to find out what is the true happiness is for each of us.

Do so by wishing the most beautiful things you might think of to your reflection in the mirror, to your family members, to those who present with you at the same time and those who have passed, to your friends and opponents.

Joining the charity, helping those in need, wishing best things to all of us, since we all are the One, is also a very helpful step.

So isn't that a proper, superior science, the essence of which we have to learn through our lives and teach that to our children? Isn't that what all religions teaching us? The initial purpose of every religion is to teach us to think in right way and to fear our own thoughts, when they are negative. God has created us by his imaging, and we have the imagination gift, too. The thought is the quantum that conducts the imagination, and conducted by the imagination. From there come our wishes and actions.

So make the world a better place: truly wish your best feeling to everyone you encounter, create constructive vibrations. When you don't want to wish your good wishes to your enemies, wish them to your beloved ones. When you don't want to see anyone, then think about the most beautiful features of the God, because every of them is true, and all of them are only the truth.

I suggest you to read a brilliant book of Deepak Chopra "Creating an affluence" The author invites the reader to pay attention to 25 features of the quantum field, by repeating one feature every day. I have practiced it, and that did work for me.

But as a natural process, I felt I want to evolve even more and when I discovered that dreaming of better things for others and wishing the best to everyone I know and every stranger I meet, makes my own life even better, I'm tuned on that iwishyou attitude, too.

I'm not suggesting you to start to talk nicely to every stranger, or to make any action towards the strangers you're seeing around. I'm talking about your thoughts and sincere wishes.

Every morning I usually drink a glass of the clean water (together with my vitamins) and creating, repeating wishes. I wish to my children, to my mother, to my every relative and everyone I encounter today, the very happiness, the very beauty, prosperity and whatever good thing comes into my mind. I also feel thankful for the every moment and everything I have. And that also works brilliantly for me too. I have encountered even more freedom, greater success in my personal life, relationships, in business. I became capable to notice things from the past, the present and their projections in the future. Analyzing the past events using my iwishyu attitude helped me to understands things I wouldn't even think about and let them go, and assume how that may affect my future and to change my thoughts and behavior accordingly … That is very important, specially for those who grow own children.

We are approaching the world where the subtle forces mean more than they have been used to and the role of the science is immense in that.

Through the research we now aware about the role and importance of our attitude, state of mind in our battle with the disease, with poverty, with depression.

In order to grow, to evolve, to acquire wealth and prosperity we have to accept that our subtle power is more essential then amount of money we own, the property we own and even our physical health. Because all can be managed by those subtle powers that are much mighty then material ones and physically are more powerful forces and facilities. But since all those subtle forces obey certain physical laws we have to understand their basics and not only obey them but use them in order to flourish, as that is the purpose of existence of the whole Universe.

Start to create beautiful tunes for others to experience, to receive, to reflect the more beautiful feedback of your affluent thoughts, dive into waves of the world, were everything naturally wonderful and is possible.

Join the club!

The books, inspired me : The Qoran, Book of the King David, The Bible.

I recommend you to read:

"The Undivided Universe" David Bohm & B.J. Hiley

"The Quantum Doctor" – A. Goswamy, PhD,

"Creating an affluence", "The Seven Spiritual Laws of Success" - Deepak Chopra.

"Breaking the habit of being yourself" Dr Joe Dispensa

"The Secret" Part 1 and Part 2.- Rhonda Byrne, or watch the movie Watch the movie "Solar Revolution" by Dieter Broers 2013

All available at online stores.

www.ingramcontent.com/pod-product-compliance
Lightning Source LLC
Chambersburg PA
CBHW050358290526
45786CB00003B/1037